EINSTEIN'S
TUTOR

$(\theta_\nu^3 su)$	$(\theta su)^2$	
$(Ksu)^3, s_j,$ $s_\vartheta (\theta s' u)^4$ $s_\nu (asu)^2$ $a_\nu (asu)^2$	(Hsu) $((\vartheta \nu x)^2, s, u)^3$ $(\theta_\nu (\vartheta \nu x), s, u)^2$ $(\theta_\nu^2 su)$	$((j\nu x)^3, s$ $(aHu)^2 s$ $(a_\eta su)^2$
$(a_\eta su)^3$ $(aHu), s, u)^2$ $(aHu)^2, s, u)$	$(Ksu)^3$	$s_j (asu)$ $((\vartheta \nu x), s, u)^2$ $(\theta_\nu su)$
$s_j (asu)^2$ $s_k (Ks^2 u)^5$ $\vartheta \nu x), s, u)^3$ $(\theta_\nu su)^2$	$(j\nu x) s_\nu$ $((j\nu x)^2, s, u)$ $((aHu), s, u)^2$ $((aHu)^2, s, u)$	$((\vartheta \eta x), s, u)^3$ $(\theta_\eta su)^2$
$(Lsu)^2$ $(j\nu x), s, u)^2$ $(\theta Hu), s, u)^2$ $(\theta Hu), s, u)^2$ $(\theta Hu)^2, s, u)$	$(a_l su)^3$	$(K_\nu s$
	$(a_\nu (j\nu x), s$	

$((a\,Hu)\,H_j,s,u$
$((\theta\,Lu)^2,s,u)^2$
$(\theta_v\cdot H,s,u)^4$

$(\theta_\eta\,(\theta\,Hu)^2\cdot H;s,$

EINSTEIN'S
TUTOR

THE STORY OF EMMY NOETHER AND THE
INVENTION OF MODERN PHYSICS

LEE PHILLIPS

PUBLICAFFAIRS

New York

PublicAffairs
Hachette Book Group
1290 Avenue of the Americas, New York, NY 10104
www.publicaffairsbooks.com
@Public_Affairs

Printed in the United States of America

First Edition: September 2024

Published by PublicAffairs, an imprint of Hachette Book Group, Inc. The PublicAffairs name and logo is a registered trademark of the Hachette Book Group.

The Hachette Speakers Bureau provides a wide range of authors for speaking events. To find out more, go to www.hachettespeakersbureau.com or email HachetteSpeakers@hbgusa.com.

PublicAffairs books may be purchased in bulk for business, educational, or promotional use. For more information, please contact your local bookseller or the Hachette Book Group Special Markets Department at special.markets@hbgusa.com.

The publisher is not responsible for websites (or their content)
that are not owned by the publisher.

Print book interior design by Sheryl Kober

Library of Congress Cataloging-in-Publication Data
Names: Phillips, Lee (Computational physicist), author.
Title: Einstein's tutor : the story of Emmy Noether and the invention of modern physics / Lee Phillips.
Description: First edition. | New York : PublicAffairs, 2024. | Includes bibliographical references and index.
Identifiers: LCCN 2024001727 | ISBN 9781541702950 (hardcover) | ISBN 9781541702974 (ebook)
Subjects: LCSH: Noether, Emmy, 1882–1935. | Mathematicians—Germany—Biography. | Women mathematicians—Germany—Biography. | Algebra.
Classification: LCC QA29.N6 P45 2024 | DDC 510.92 [B]—dc23/eng/20240130
LC record available at https://lccn.loc.gov/2024001727

ISBNs: 9781541702950 (hardcover), 9781541702974 (ebook)
LSC-C
Printing 1, 2024

To my children.

Contents

Introduction   1
"The single most profound result in all of physics"

1 Intersecting Paths   11
"The most beautiful of all existing physical theories"

2 Gravity   53
"I have hardly come to know the wretchedness of mankind better than as a result of this theory."

3 The Theorem   99
"It would not have harmed the Göttingen old guard to have been sent to Miss Noether for schooling."

4 Emmy Noether Gets a Job   145
"How easy this decision would be for us if this were a man."

5 The Purge   163
"Mathematics knows no races or geographic boundaries."

6 Emmy Noether in America   175
"Mathematicians, like cows in the dark, all look alike to me."

7 Reawakening   217
"How can it be that mathematics . . . is so admirably adapted to the objects of reality?"

8 Legacy   241
"If she knew how useful her mathematics had become today, she'd probably turn over in her grave."

Contents

Appendix   265
"A little lower layer" —*Captain Ahab*

Acknowledgments   311

Notes   313

Index   335

Introduction

"The single most profound result in all of physics"

Whether you're a scientist, a student, or someone else with a general interest in the history of ideas, you certainly have some notion about what physics is and how it developed over the centuries. Whatever the origins of these notions, chances are that the name Emmy Noether doesn't spring immediately to mind. Albert Einstein, Erwin Schrödinger, Paul Dirac, Niels Bohr, Werner Heisenberg . . . these are some of the names familiar to anyone who's read books about the invention of modern physics—about the momentous discoveries of the twentieth century that changed our conceptions of time and space, and of the nature of reality itself. And these are the names that figure prominently in the textbooks used by physics students.

Who, then, was Emmy Noether?

In the pages ahead I hope to convince you that she deserves a place alongside the names mentioned above in the history of physics—and of science in general—and that this is due to the impact of discoveries that she published in 1918.

Her four closely connected results, collectively now called Noether's theorem, supply the foundation of the present-day search for the holy grail of physics: a unified theory that will join quantum mechanics with gravity. Noether's theorem has also supplied the methodology for constructing the most accurate theory in the history of physics: the standard model. This framework encompasses all the elementary particles and their interactions—it is our modern theory of matter. In addition, Noether's discovery solved a persistent mystery in Einstein's recently completed general theory of relativity: the conundrum of energy conservation, a problem whose solution had eluded Einstein and several of the greatest mathematicians in the world. Attacking this puzzle began the chain of reasoning that eventually led Noether to her great theorem. Along the way, she wound up tutoring Einstein in some of the math he needed to complete his work, thereby becoming one of several uncredited authors of what remains our current theory of gravity. I'll have much more to say about unification, the standard model, general relativity, and their connections with Noether's theorem throughout the book. For now, it's enough to note that these subjects essentially define the principal concerns of what we think of as fundamental modern physics, and they are all intimately connected with, and often depend on, Noether's theorem.

The theorem goes beyond providing the foundation of current physical theory and supplying the touchstone for the physics of the future. It provides the modern definition of the concept of energy and clarifies the importance of symmetry in nature. It brings order to the physics of the past, completing its theoretical structures and rendering it whole. The theorem's revelation of the active role of symmetry throughout nature is so provocative that it is now being exploited in biology, computation, economics, and elsewhere.

In this book, I trace the surprising trajectory of the theorem: its origins and the events and lives that created the conditions for its birth. I explore how the unique genius of Emmy Noether allowed her to see something completely unexpected and arrive at insights that still astonish those who encounter them a century later. I'll recount how the theorem became dormant for decades and was almost lost to the world, and how it was rediscovered by physicists creating a new theory of matter. Finally, we'll see how Noether's theorem has now gained a new life, guiding research in fields far removed from physics.

Emmy Noether's story is that of a woman pursuing her passion for over three decades in a world offended by her simple intention to be a mathematician. But that is what she would be, however she could. Despite being denied opportunities, passed over, excluded, and expected to work without pay or position, she not only persisted but surpassed the men around her. Her story is also a story of solidarity and the loyalty of those few men who championed her cause.

This view of Noether's importance is bolstered by the judgments of many of our leading physicists, including Nobel Prize winners such as Frank Wilczek. Wilczek and other top researchers, intimate with how their science works and its debt to Noether's theorem, have occasionally stepped outside their offices and laboratories to write about science for the layperson, explaining the big ideas in physics, its history, and its possible future development. Along with me, they believe that Noether's theorem is one of these big ideas, and they consider its creator a neglected figure in the history of science. According to Wilczek, Noether's theorem is "the single most profound result in all of physics."[1]

Renowned physicists Leon Lederman and Christopher Hill say that Noether's theorem is "one of the most important mathematical theorems ever proved in guiding the development of modern physics, possibly on a par with the Pythagorean theorem" and that it "rules modern scientific methodology."[2] (Lederman is a Nobel Prize winner and coined the term *God particle*.)

Brian Greene, a theoretical physicist well known for explaining modern physics through books and television and a leader in current attempts at building a unified theory, believes that "Emmy Noether's theorem is so vital to physics that she deserves to be as well known as Einstein. Yet, many have never even heard of her."[3]

Emmy Noether was born in Germany in the late nineteenth century and died in the United States in the twentieth. She devoted her life to research and teaching in pure mathematics. She did so out of unquenchable passion for math, despite a cruel series of obstacles and injustices placed in her path for one reason only: that she was a woman.

She sought to study math in the university, but women were not allowed to enroll—so she audited, when permitted. After German society relented and allowed women to matriculate, she took her degree. She went to work at the premier university for mathematics in the world, by invitation of the greatest mathematician in the world—but without pay or position, because women were not allowed to be teachers. After about half a decade, that rule finally changed, too, and she was hired—but reluctantly, with minuscule salary, expressly excluded from any civil service benefits and warned that she had no authority.

She was one of the first victims of the Nazi purges of Germany's faculty rolls, as she doubly offended by being both female and Jewish. Escaping to the United States, as her colleague Albert Einstein had already done, she was placed not alongside him and other refugee members of the intelligentsia at the new Institute for Advanced Study but in an obscure position, deliberately stowed at a distance from her male peers.

If the sound of Noether's voice could somehow reach us one hundred years after she lived, we would hear no wail of complaint. Instead, we would hear the loud laughter so often described by her compatriots. She seems never to have voiced any objection to her lot or struggled to improve her position, although she was energetic in fighting for her friends. She simply proceeded to do whatever the circumstances permitted so that she could carry out her interests—her only interests—which were to discover and teach mathematics.

Denied the right to be a student, she audited; denied the right to a job, she worked without pay; forbidden by Hitler's regime to teach at all, she taught in secret, in her apartment. Throughout it all, she enjoyed life, she raised the spirits of everyone around her, and, from joy, irony, and the delectation of absurdity, she laughed.

———

The full answer to the question of why Noether is so little known, given that she contributed as much to shape the history of physics over the past century as did any other person, is complex, sometimes subtle, and multifaceted. I will answer this question gradually throughout the book; in this introduction, I'll give an outline of some of the reasons.

Noether's relative obscurity is due, first of all, to the systematic suppression of her reputation and place in history by colleagues, officials, and scholars who minimized her contributions and importance. The evidence that this minimization was due to her sex is stark and abundant, sometimes taking the form of bald statements by the perpetrators themselves, who knew no reason to disguise their prejudice. In matters relating to her circumstances during her life, we can add to this the fact that she was a Jew in Germany in the 1930s.

For her part, Noether did nothing aimed at bolstering her own prominence or position. She was supremely generous, helping colleagues and students in the early stages of their careers by making gifts to them of mathematical results: theorems that she had proven and problems that she had solved but hadn't bothered to publish. She encouraged these young colleagues to refine or extend her work and present it to the world as wholly their own. These were valuable gems that a normal academic type would have jealously guarded and polished until they were ripe for a career-boosting publication. But Noether, who was overflowing with ideas and results, was happy to pass many of them along for the benefit of her friends. Although aware of the importance of her own work in mathematics, she did not promote herself, and she rarely referred to her monumental result in physics, which was for her a side issue that she forgot about almost as soon as she had created it. This habit, combined with the casual attitude, even by her supporters, to issues of credit and priority, conspired to cause her role in her contributions to be overlooked, even after they had become part of the working machinery of physics.

In recent decades, we've seen a gradual rehabilitation of Noether's reputation and a growing recognition of her theorem as a crucial component in the development of fundamental physics since 1918. As mentioned, this revision is largely carried out by scientist-writers who are aware that theory building rests on Noether's theorem as a foundation, as an omnipresent guide and constraint. I intend this book to be a part of this revision, and I am confident that, before too long, the idea of a book chronicling the story of modern physics without Emmy Noether as a main character will be as unthinkable as one that forgot to mention Einstein.

How did I become aware of Emmy Noether and what she gave to physics?

During my college days, while browsing through an advanced textbook on classical mechanics, I came across something that stopped me in my tracks. A routine development of the subject had taken a minor detour to present some results that were surprising and that struck me as deeply profound—as a demonstration of how physics had the ability to capture the harmony and unity of nature.

What I had come across in the textbook were proofs that the conservation laws of classical physics, the familiar laws of the conservation of energy, momentum, and angular momentum, were each equivalent to a symmetry in time or space. Ideas like the conservation of energy were not just additions to mechanics that made it easier to solve problems involving, say, the trajectory of cannonballs; they were implied by the very structure of space and time. Such a connection among concepts that had previously seemed unrelated, demonstrated by unambiguous mathematics, not only was unexpected but also seemed to have implications that reached beyond physics and toward philosophy. At the very least, these beautiful connections suggested a myriad of questions and pushed me to look deeper into my subject.

These relationships stuck with me after college, all through graduate school. But I had gone all the way to a PhD without ever hearing about such things again, at least not directly. At the time, this omission didn't seem strange. I assumed that the results were confined to classical mechanics, although they remained in the back of my mind throughout my studies.

Many years later, I convinced the editors of an online science magazine to let me write a popular article about the connections between conservation laws and symmetry. I still didn't know much about the subject, but I knew it was intriguing and important, and not widely understood. I was sure that I could also make it interesting to nonspecialists, to show how physics could be provocative and fascinating apart from the wonders of the quantum world and relativity—to show how even classical mechanics could be beautiful.

In doing research for that article, I finally learned where these ideas had come from. They were simple, special cases of a deep theorem published in 1918 by someone I'd never heard of, someone whose name had not been mentioned once in my many years of physics training. Her name was Emmy Noether, and her result was called *Noether's theorem* by the physicists who knew about it. In fact, those physicists seemed to belong to a secret club. They spoke of this theorem as one of the, if not *the*, single most important result in theoretical physics. They bemoaned the fact that its discoverer was not more widely known—was barely known at all, rarely mentioned either in classrooms or in popular histories of science. And these scientists were neither crackpots nor cultists. They were some of the biggest names in physics.

I kept reading about the theorem and the person behind it. I learned the amazing, inspiring, and tragic story of her life. I learned far more about some of the people whose lives had intersected hers, whose names I *had* encountered in my studies: David Hilbert, Felix Klein, Hermann Weyl, Einstein, and others. I went further, delving into archival materials that revealed the untold parts of the story about her brief time in the United States. I learned that I had, quite literally, walked on her grave without knowing it.

You don't have to know any physics or advanced mathematics to read and understand this book. If you have some vague memory of what the Pythagorean theorem is, that should be plenty. I intend to describe the meaning and content of Noether's theorem in a way that can be understood and appreciated by everyone. This is possible, even though the theorem requires some advanced mathematics to prove, because it possesses an intuitive core. Its essential meaning for physics and beyond is thoroughly explainable through prose. I will guide you through an honest understanding of the meaning of this result, so that you'll be able to appreciate its more recent uses in other fields, outside of physics, such as in biology and economics. If you follow where I lead you, you'll see how this one powerful idea unifies many realms of thought that seem superficially unrelated.

This is not a biography of Emmy Noether; rather, it's the biography of an idea. She devoted her life to mathematics, and a faithful account of her life must stay close to the work that was important to her and its influence on the history of thought. To this end, I will take the time to dissect various topics in physics and mathematics as honestly as I can without resorting to equations, to relate them to the idea that will remain at the center of the story: Noether's theorem. Much of this dissection is relegated to an appendix. There the interested reader will find a deeper layer of detail, where the history of, and relations between, ideas in physics and mathematics, and their connections to Noether's thought, are explored. The somewhat more technical nature of the appendix will satisfy the more committed or mathematically inclined reader, while the sequestration of its details allows the other aspects of the story, namely, its intertwining, interpersonal threads, to proceed with more directness.

———

For those who want to delve deeper into anything that I mention, I supply many references to buttress my claims or to suggest further reading. Some of the references are to the technical literature, to serve the purpose of perhaps nonobvious sources for specialists; other references range from popular articles and books to videos and cartoons. In no case is a reference an endorsement. Many of the sources I cite contain errors (I've lost count of how many biographical notes about Noether state erroneously that she died of cancer), but aside from the mistakes, the sources contain interesting takes or information. You'll notice many books, such as Constance Reid's excellent biography of Hilbert, referred to repeatedly. Take this to stand in for the conventional suggestions for further reading.

———

I am not trained as a historian, yet to tell this story, I had to try to become one. I now have a new appreciation of the complexity and melancholy of the historian's task. One wants to tell a story of the past, where each event leads naturally to a subsequent event, where people's motivations are comprehensible, and where things hang together in a way that at least seems something other than utterly random and chaotic. Yet for each important event or turning point, there are contradictory accounts, lies, and fanciful imaginings in the sources that we dig through to try to construct a version of the past. Motivations are obscure. But somehow, one must make out of this stew of confounding details some kind of story, or else there would be nothing to call history. I find myself nodding knowingly at the words that Mark Twain put in Herodotus's mouth: "Very few things happen at the right time, and the rest do not happen at all. The conscientious historian will correct this defect."

There is also the need to constantly fight off the tendency to judge or understand people of the past as if they were our neighbors in odd outfits. The past is another world, cut off from our ideas no less than a geographically isolated region would be. We must aspire to be anthropologists of time.

From all this comes the complexity.

The melancholy arises from this: after a while we get close to the men and women of the past, with whom we have spent so much quality time. We care about them as if they were friends or relatives. And we are burdened with the bittersweet gift of hindsight. When we reconstruct the crossroads faced by our heroes, there are times when we know they are about to take the wrong path, which will lead them to misery. Or perhaps there is only one road to take, and as they march along it, we know that they are unprepared for the terror that awaits. In neither case can we warn them; there is nothing we can do to help. But perhaps if we can allow our small, irrational pangs of anguish at these times to animate our stories of the past, they will resonate more meaningfully for the reader. It is a consolation I allow myself, at any rate.

Finally, in getting to know these characters from history, I had a pleasant surprise. When we are young, we tend to look for heroes, in the present or in the past. But when we find them, and then look deeper into their lives, we're almost always disappointed, or even horrified. They fail to live up to our standards. We become cynical. As I studied the main characters of my story, and as they seemed of reliably sterling character, therefore, I kept waiting for the other shoe to drop. Yet it never did. Our principal players, especially Emmy Noether and David Hilbert, did not disappoint. They were courageous, generous, and brilliant at every turn. I might even dare to say that they can be heroes for some of us.

1

Intersecting Paths

"The most beautiful of all existing physical theories"

In this chapter we'll meet the three people who play the central roles in our story.

First, of course, is Emmy Noether, around whose life and work everything else revolves. We'll follow her life from late adolescence until 1915, when her path intersects that of Albert Einstein.

I'll trace the relevant details of Einstein's life and work during this same period. There's a bit less to say about him in this chapter, as he becomes a more active participant after 1915, with the events portrayed in the chapter that follows. But it's good to know how he enters the picture, and why.

Finally, there is David Hilbert. Acquainted with Noether's family for many years, acquainted as well with the work of Einstein, and eager to know more, it is Hilbert who brings the three together, who creates the conditions for Noether to make her momentous discovery.

Aside from these three most indispensable people, several who performed supporting roles make their first appearance in this chapter. One is not a person, but a place. I'm not the first who has written about how the spirit of the great university at Göttingen seemed to play an active role in the long series of momentous discoveries in mathematics and science that took place on its stage. In its traditions of freedom, tolerance, and intense meritocracy, Göttingen acted as a silent colleague working alongside and inspiring a long parade of leaders in the history of thought over the last three hundred years.

There is also Felix Klein, another giant of twentieth-century mathematics. He was an educational innovator who, as part of his program to prepare the Göttingen math department for a glorious future, schemed to bring Hilbert under its wing.

These are the events and developments that, beginning around the turn of the last century, eventually led Noether, Hilbert, and Einstein to find themselves in the same place in the summer of 1915.

Emmy

The boys didn't want to dance with her.

In 1890s Germany there were no forms of electronic entertainment. People often socialized by gathering in each other's houses, playing musical instruments, and dancing. Emmy Noether's mother was good at the piano and often played with an accomplished violinist.

By all accounts Emmy was a bright, lively, kind, and friendly girl. But here we run into the same difficulty as others who have been moved to write about her life. As her circumstances were not particularly notable, as nobody around her could possibly have foreseen the Olympian position she would occupy in the history of science, there was no particular reason to record the details of her early years. Such is the case with every figure who, beginning as an inconspicuous caterpillar and suffering an obscure pupation, bursts forth as a rare, stunning butterfly. And those who gasp at this sunlit, fluttering audaciousness will again fail to note the camouflaged brethren munching on the leaves nearby. This state of affairs, a common-enough paradox facing the biographer of the scientist or artist, may have been made a bit worse in this case because our subject, after all, was only a girl.

The mothers of the teenage boys who happened to be among the guests at the Noether family's large second-floor apartment or at other social venues knew that she loved to dance, so they would cajole their sons to give her a chance now and then.[1] After all, a book of etiquette published a few years before Emmy's birth did have this to say: "A gentleman of genuine politeness will not give all his time and attention to the belles of the evening, but will at least devote a little thought to the wall-flowers who sit forlorn and unattended, and who, but for him, might have no opportunity to dance."[2]

Nobody had anything against her. In fact, everyone liked Emmy. But she was not very pretty or at all graceful. She was nearsighted and spoke with a bit of a lisp. The boys had their eyes on the other young ladies.

There may have been something else keeping the boys away. The gatherings at the Noether house had an academic flavor. The family lived in Erlangen, a university town. The University of Erlangen was one of Germany's free universities, called such because it was independent of any church. Emmy's father, Max Noether, was a prominent mathematics professor there. He was not a healthy man, having suffered from polio as a boy and been left permanently affected.

Naturally, many of the guests were university types. Now and then, a challenge was issued to the assembled adolescents, sometimes in the form of a math puzzle: "Who can tell me . . . ?" Emmy would supply the answers right away, while the others were still trying to figure out what the questions meant.

We live in a different world now, but some things haven't changed much. Many men, in the presence of women demonstrably more intelligent than they are, feel . . . emasculated. This fact of life is easy enough to observe in our egalitarian moment in history. In the late nineteenth century, Germany was not merely comfortably patriarchal. It was distinctly behind most of the rest of Europe, socially and legally, when it came to rights for women. We can imagine just how uncomfortable the naturally self-conscious adolescent male was made to feel when outshone by a member of the category of person that, it was universally known, was simply not suited to intellectual work.

No, they didn't want to dance with her. They knew they would not fare well in a dance of wits.

Twenty years later, someone else would set a puzzle before Emmy. The most famous mathematician in the world, David Hilbert, would ask her a question. Her answer would reverberate through all of science. It would unify physics and answer many of its questions, guiding its future development down to the present day. But we're getting a bit ahead of ourselves.

Göttingen, 1890

Göttingen is another university town, about two hundred miles north and slightly west of Erlangen along modern highways. The university, also free, shared with Erlangen a strong reputation in the sciences and mathematics.

In 1890, the mathematician Felix Klein had command of the university's math department. Today, Klein's name is well known in physics circles and among mathematicians. He's famous even among laypersons because of his wide-ranging creativity in many areas of math and science, including the devising of such recreational curiosities as the Klein bottle.[3] This is a three-dimensional version of the Möbius strip that some of us were encouraged to construct in primary school by making a paradoxically single-sided loop out of paper. Klein's bottle extends the paradox into the third dimension, creating a container that has, rather than the customary inside and outside, one continuous side. They make lovely conversation pieces.

Klein was devoted to the life of the mind, but he could play rough when it was called for. The mathematician was more than willing to expend considerable energy and time in political horse trading to get what he wanted for his department and university. He had built up the math department into an international powerhouse, extracting money and a series of concessions from the German bureaucracy.

Shortly after accepting a professorship at Göttingen, Klein had embarked upon an extremely ambitious plan to transform the university into a kind of think tank for the physical and mathematical sciences. His ambitions extended beyond Göttingen to science and technological education in Germany overall, including secondary and grammar school policies. To these ends, Klein lobbied for the admission of female students to Göttingen and tried to attract more foreign students as well.[4] As an educational administrator, Felix Klein was ahead of his time, with an amazingly comprehensive vision for science education in general and Göttingen's place in it in particular. As an example of his original and forward thinking, Klein tried, with some eventual success, to raise funds from German industries to fund technologically oriented education and research. He took advantage of the industrialists' appreciation of their continuing future needs for a population of workers literate in various applied sciences. This approach has only recently been echoed in the United States, with corporate sponsorships occasionally filling the gaps in the government funding of public education, as Klein intended for Göttingen. Klein also greatly increased the activity in the applied sciences in Göttingen, believing that these fields, along with the pure physical sciences and mathematics, were strongly connected into a unified intellectual whole and could support and strengthen each other.

Klein did not accomplish these visionary administrative feats without some help. His accomplice in the government, a man named Friedrich Althoff, had authority over all higher education in Prussia. Althoff was sufficiently sympathetic with Klein's ideas to do a certain amount of battle in their support with the bureaucracy. Althoff's task was aided by the bluntness of his personality. The quantum mechanics pioneer Max Born remembered him years later as "well known and feared for his lack of consideration and rudeness."[5] It certainly didn't hurt that Klein and Althoff were good friends; they had been in the army together during the Franco-Prussian War.[6]

A brief look at the profound changes brought to Europe by this one-year conflict helps reveal the cultural and political landscape inhabited by our protagonists. The remaining independent German states found themselves part of a unified Germany at the war's end in May 1871. Nevertheless, they retained much of their separate cultural identities and a certain amount of administrative autonomy. For example, Althoff was in charge of educational policy for the Prussian state only. But since Prussia was the dominant force in unified Germany, and as the most important universities were in Prussia, Althoff occupied the most important policy-setting position for German education in general.

Our three protagonists came from various regions of the recently unified Germany. Hilbert's home town was in the eastern outskirts of Prussia, the state where he pursued his career as well. Einstein was born in the town of Ulm (whose motto is "The people of Ulm are mathematicians") in the southern German kingdom of Württemberg. Erlangen, where Emmy Noether was born, raised, and educated, is in Bavaria. The Franco-Prussian War led to the downfall of Napoleon III, the establishment of the Third French Republic, and a profound reduction of French power in Europe, along with a major loss of territory to Germany. The European reconfiguration was widespread, including the unification of Italy.

This chapter begins in the aftermath of the Franco-Prussian War and ends at the early stages of the Great War, or what we now call World War I. The Franco-Prussian War was part of the impetus for the Great War, just as World War I was one cause of World War II. Among other motivations, the glorious outcome of the Franco-Prussian War inspired a kind of militaristic pride in a large portion of the German people and rendered additional campaigns palatable. As we'll see, many of the principal actors in this book recoiled against this strain of militarism.

The death of a key mathematics professor in another university initiated a typical round of negotiation and German academic musical chairs. In 1894, a rare spot opened up in the math department at Göttingen, giving Klein an equally rare opportunity to mold the department further in the direction of his ideas and to boost its international reputation even higher with the addition of a rising star with a shining future.

Klein knew exactly whom he wanted. He sent a letter marked "Very Confidential" to his top choice to fill the spot. He didn't know if he would be able to arrange an official offer to his candidate. It was a complex game that involved a handful of universities and was watched over at the ministerial level.

The scheming around faculty appointments was one of several facets peculiar to the German university system, a stew of ancient traditions that various authors have credited for that nation's intimidating position in the world of scholarship over the centuries. One aspect that made the appointment of particular professors critical to the growth and success of their host departments is related to a kind of freedom enjoyed by German students. This tradition is unknown in most other countries, for example, in the United States. A university student in Germany was free to take any classes he desired and could even travel from university to university to attend the lectures of any professor whose reputation attracted his attention. Part of an institution's income, and the personal income of lecturers in the lower ranks, depended on attracting students.

Klein needed to know whether he would be wasting time and political capital, so he insisted on one promise from Hilbert: that if the offer were extended, his target would accept.

Hilbert, 1890

David Hilbert didn't have to think very hard to formulate a reply to Klein's "Very Confidential" letter. He didn't negotiate. He sent off his answer right away: "Without any doubt I would accept a call to Göttingen with great joy and without hesitation."[7]

Hilbert was a thirty-two-year-old math teacher at the University of Königsberg. You won't find Hilbert's place of birth on a modern map—at least, not with the name it had back then. Königsberg, in eastern Prussia, was absorbed into Russia at the close of World War II and was renamed Kaliningrad. Any Königsbergs that you find in a current map of Germany are not Hilbert's Königsberg. But the famous mathematical puzzle associated with the town retains its original name. The so-called Königsberg bridge problem asks if you can find a path that crosses each of its seven bridges once and only once.[8]

Königsberg looms rather large in the history of the intellect; the town gave birth to or nurtured several important mathematicians and scientists and is the cradle of Immanuel Kant, whose writings are thought to have influenced Hilbert's philosophy of mathematics. (Hilbert's influential book, *The Foundations of Geometry*, begins with a quotation from Kant: "All human knowledge begins with intuitions, thence passes to concepts and ends with ideas."[9]) German university students of this era were in the habit of traveling from school to school, sampling the wares of several institutions on their way to a degree. Hilbert, however, stayed in his home town to attend the university, where he became good and lifelong friends with Hermann Minkowski, who was eventually to become famous in a generous portion of four-dimensional space-time, and who will make several appearances in later chapters.[10] Hilbert had turned out a series of stunning results in mathematics at Königsberg; when Klein's invitation reached him, the young professor was still building his career, but some of these results had already attracted worldwide attention.

During this time, Hilbert formed his habit of carrying on mathematical discussions while walking around outside, evincing a relative distaste for the more conventional environs of offices and libraries. While this preference contributed to the overall impression of eccentricity or, perhaps better said, strong individuality that Hilbert already projected, this restless habit was not unheard of or unique among German mathematicians. Hilbert's eventual academic home, Göttingen, had somewhat of a tradition of peripatetic mathematics, encouraged by the inviting woods that surrounded it (largely preserved today). The custom of mathematical strolls would be happily joined in by his colleagues there, including Emmy Noether, who, as we'll see, eventually transported this method of working to the United States. Hilbert and Minkowski used to go for regular walks with a favorite mathematics professor, Adolf Hurwitz.[11] They were soon joined by others, and the walks became a daily moveable math seminar, where much knowledge was transmitted and new knowledge created.

Back at Göttingen, Felix Klein was happy to get Hilbert's acceptance, but now he knew he had a real battle ahead of him. The Faculty Senate had to sign on to a hiring decision such as this.

Hilbert was a hard sell because of his reputation for, among other things, cultivating an overt contempt for any type of authority. The young professor, toiling at his relatively small-town university, was already notorious.

Scandalous stories had emanated from Königsberg. Hilbert was known to be shockingly casual in dress and manner for the era. He was a regular at dances and other social functions, where he shamelessly flirted with handfuls of young women. He did not fit many people's image of a proper German university professor.

Hilbert's champions in the Göttingen Faculty Senate were the leaders of the mathematics contingent, famous mathematicians in their own right, and the scientists. They knew all about his peculiarities, but they didn't care. They were eager to capture Hilbert, whose attachment to Göttingen would greatly enhance the already-stellar reputation of their group.

Their opponents were the professors of philosophy, philology, literature, and theology, who were far more conservative than the scientists and mathematicians. This is a theme that we will see again. These old men of Göttingen wanted nothing to do with Hilbert. They needed careful handling.

Klein prevailed in the end because of the significant clout he had as a legendary mathematician and teacher, helped by his connection with Friedrich Althoff, the minister of education. During his years building up his department, Klein had also become a master academic strategist, skilled at negotiating with and cajoling his colleagues. At one point, another faculty member, apparently not closely familiar with Hilbert's reputation but aware of his youth, chided Klein that he seemed to be looking for an easygoing candidate, perhaps thinking that Klein preferred somebody that he could push around. Klein reassured his colleague that, to the contrary, "I have asked the most difficult person of all."[12]

Klein was in a unique position to manipulate the humanities professors. You could say that he had pursued a second career in educational administration, having become unusually adept in the intricate political maneuvering peculiar to German academic life.

After a series of tense meetings and negotiations, Klein succeeded. Hilbert was in.

Emmy Noether, Mathematician

It was 1900, and Emmy Noether had just turned eighteen. It was looking likely that she would never be married.

Consequently, she followed a conventional path for a smart young woman of her time and class. The teaching of languages was one of the few socially acceptable quasi-academic callings available to females, and for an unmarried young woman who was part of an academic family, this occupation was nearly inevitable. She took the required examinations and became certified to teach French and English to girls.[13]

She never did. Instead, Noether yielded to the restless stirrings of an unsatisfied desire to gain knowledge beyond what her conventional schooling had provided. We do not know what set her on this course. Perhaps she had caught glimpses of the arcane scholarship in which some of her relatives and their friends and colleagues were immersed. Her father, who labored near the edge of current mathematical knowledge ("one of the finest mathematicians of the nineteenth century," according to some), might have had the strongest influence.[14] Before long, Noether the daughter sought to scratch this itch by auditing courses in a variety of subjects at the university.

How unusual her educational pursuit was can be seen from one statistic: there were 984 male students at Erlangen and *two* female auditors.[15] One of the obstacles was the hostility of some of the professors to the mere presence of women in their classrooms. The Ministry of Education had been battling with the conservative professoriat for some time to impress on them the rule that they must not exclude women for arbitrary reasons. This battle would continue even after Noether finished her formal education. The position of Emmy's father helped convince most of the professors to allow her to audit, but she did have to fight for this right on occasion. Women were, however, now allowed to take graduation examinations, if not the examinations for individual classes, and in 1903 Noether took and passed the graduation exams.

By now the mathematics bug had bitten her hard. She continued to audit advanced classes, and not only at her home university: she spent a semester at Göttingen, packing a great many lectures into her auditing schedule. It must have been a heady experience for an aspiring mathematician. The list of professors whose classes she sat in on reads like a roll call of math celebrities, including many names that, like Klein and Hilbert, would be familiar to any student of mathematics or physics today.

Looking back on this time, the great mathematician and physicist Hermann Weyl would describe it this way: "Emmy Noether took part in the housework as a young girl, dusted and cooked, and went to dances, and it seems her life would have been that of an ordinary woman had it not happened that just about that time it became possible in Germany for a girl to enter on a scientific career without meeting any too marked resistance. There was nothing rebellious in her nature; she was willing to accept conditions as they were. But now she became a mathematician."[16]

It's possible that Weyl, given a chance to edit these remarks, might have rephrased his "any too marked resistance." He became good-enough friends with Noether to be unaware of the resistance, marked indeed, that she faced at every stage of her attempts to pursue her scientific career—obstacles that did not exist for him and other men who shared her qualifications. He was keenly aware of her talents as a mathematician and considered them superior to his own. As detailed later in the book, his other statements and actions suggested that he considered her treatment unfair. Weyl wants us to understand his friend Emmy Noether's attitude and state of mind: that the course of her life was fueled by her overwhelming passion for mathematics. Anything unconventional in the path she chose was a result of her following that passion; without it, there would have been nothing to motivate her to deviate from the road more traveled.

Hilbert at Göttingen

At Göttingen, David Hilbert lived up to the best hopes of those who had wanted him there, and confirmed the worst misgivings of those who had been against him.

Once installed at the university, Hilbert began to emboss a series of indelible memories on many of the people who encountered him. One student later recalled the "strange impression" that Hilbert made—that of a quick and casually attired man who "did not look at all like a professor."[17]

For Hilbert's part, he found the atmosphere at Göttingen somewhat too formal for his tastes. There were plenty of motivated students in mathematics; they had been drawn there by the international reputation of Felix Klein. But the atmosphere at Göttingen was still, at that time, rather formal and chilly, with academic ranks among the different levels of faculty and between faculty and students carefully observed.[18] This was not Hilbert's style.

Before a year was out, Hilbert had redoubled his unconventional approach to life and academia and had begun to surround himself with people whose company he enjoyed and with whom he could profitably discuss mathematics, without regard to the behavior and customs that might be expected of a man in his position.[19] The students and junior faculty were thoroughly charmed by Hilbert, who presented a marked contrast to the imposing, godlike figure of Klein. They were amused by his patterns of speech and his provincial Königsberg accent and permitted themselves a little good-natured mimicry of some of his characteristic expressions.[20]

Göttingen's students soon learned of another side to Hilbert's nature. He was intellectually rigorous and unforgiving of slovenly argument or boring presentations and could be fierce in his reactions. According to Weyl, who was one of his students from this period (and who was to become a famous mathematician himself, a chairman of the math department, and an important figure in this story), "You had better think twice before you uttered a lie or an empty phrase to him. His directness could be something to be afraid of."[21]

Weyl had joined the ranks of students at Göttingen in 1903.[22] As had so many others, he soon found himself under the spell of a kind of intelligence that he had simply never encountered before: "The doors of a new world swung open for me, and I had not sat long at Hilbert's feet before the resolution had formed itself in my young heart that I must by all means read and study whatever this man had written."[23]

Once, after Hilbert learned that one of his students had switched from math to poetry, he said that this was just as well, since the student lacked the imagination to be a mathematician.[24] And this was one of Hilbert's few general complaints about his students at Göttingen: that they sometimes failed to display sufficient *imagination*.

While at Göttingen, Hilbert redoubled his reputation for eccentricity. By the end of his first year there, he and his young family had a house built and had made certain arrangements to facilitate Hilbert's preference for working outdoors, away from dusty books and libraries. Hilbert actually had a giant blackboard attached to the wall of his neighbor's house and had a covered walkway constructed so that he could work outside even in the rain.[25] A few years later, he would acquire a bicycle and learn to ride at the age of forty-five.[26] Hilbert's new vehicle became part of his regular work routine, which involved taking turns scribbling on his outdoor blackboard, walking around, gardening, and enjoying quick spins on his bicycle. He hung out in pool halls with the junior faculty, toward whom he was expected to be aloof. In the winter, he barged into his classroom on skis. When he wanted to go for a walk with Minkowski, Hilbert would go to his friend's house and throw pebbles up at his window, another practice that failed to burnish his social reputation.[27]

Perhaps an even more telling hint of Hilbert's unconventional attitudes is the story that when his son, Franz, began his schooling, he was asked what religion he belonged to, and the boy had no idea.[28]

Throughout all the turmoil, conflict, and, finally, resounding success that marked Hilbert's career at Göttingen, his wife, Käthe Jerosch, was a constant source of support and a crucial go-between and facilitator. She understood when he needed some quiet time to work and served as a gatekeeper to the constant stream of students and colleagues who came knocking on his door. It might be fairly said of Hilbert that he was a complicated man, and nobody understood him but Frau Hilbert.

In 1900, a mere five years after joining the Göttingen faculty, Hilbert gave a speech in which he defined ten (later expanded to twenty-three) critical unsolved problems. These were not simply puzzles but a kind of blueprint for the shape of the future of math.

It was Hilbert's friend Minkowski who planted the idea for the theme for the lectures.[29] He suggested that, in planning a turn-of-the-century talk, Hilbert might consider a "look into the future" and construct a "listing of the problems on which mathematicians should try themselves during the coming century. With such a subject you could have people talking about your lectures decades later." In fact, over a century later, we are still talking about them.

Such was Hilbert's intellectual authority by this time and his sweeping command of the entire universe of mathematical research that these so-called Hilbert problems had a profound influence over the development of the field. They are known to every mathematician to this day, and some of them remain unsolved. Finding a solution to an open Hilbert problem would make a name for any mathematician who managed it.

Dr. Emmy Noether

Emmy Noether passed her graduation examinations in the summer of 1903 and immediately proceeded to postgraduate studies.[30] By now she was committed to mathematics and headed for the place that everyone now knew was the center of mathematics in the entire world: Göttingen University.

She spent a semester there, again officially as an auditor. Women were still not legally permitted to matriculate at German universities, even if they were allowed to sit for the graduation examinations. She audited lectures by a roster of geniuses whose names are permanently recorded in the history of science and mathematics: Karl Schwarzschild, Hermann Minkowski, Otto Blumenthal, and Felix Klein and David Hilbert themselves. Schwarzschild was an astronomer with preternatural mathematical skills; the others, as noted earlier, were mathematicians.

After that heady semester at Göttingen, Noether returned home. The law was finally changed to give women the same rights as men to enroll in universities and receive degrees. In the fall of 1904, she officially entered the University of Erlangen as a student of mathematics.[31] Erlangen was organized into "schools"; mathematics was Section II of the School of Philosophy. After Noether matriculated, this school contained 46 male students and her. The only other women were at the medical school, which contained 3 official female students and 2 auditors, among 159 men.

Her father, Max, shared responsibility with another notable mathematician, Paul Gordan, for teaching the main courses in the math department. Emmy often attended her father's lectures with her brother Fritz, who was studying math and physics.

Gordan was one of the world's experts on something called *invariant theory*. Emmy Noether began to study this subject intensively under his guidance, and in December 1907, she received a PhD, summa cum laude, with a thesis on the subject.

The next two chapters will set the stage for Noether's theorem and describe both the external circumstances and the internal preparation that culminated in her groundbreaking discovery for physics. A main thread of Noether's internal preparation had to do with the evolution of her mathematical style and attitude; this journey began with her tutelage under Gordan, and its first signpost is her thesis. Gordan was well known for his highly detailed, computational approach to mathematics. His papers often consisted of immense strings of equations uninterrupted with prose. Noether absorbed the approach of her research adviser, and her thesis reflects this style in the extreme. It's not unusual for researchers at the beginnings of their careers to adopt the approaches of their mentors, even those, like Noether, whose originality would soon lead them along a widely divergent path.

In Emmy Noether's case, however, this divergence was pronounced. Hermann Weyl would, much later, look back over Noether's career: "A greater contrast is hardly imaginable than between her first paper, the dissertation, and her works of maturity; for the former is an extreme example of formal computations and the latter constitute an extreme and grandiose example of conceptual axiomatic thinking in mathematics."[32]

Noether would probably have agreed with him. After she entered her more mature phase of research, she became impatient with any mention of this thesis, calling it "crap" and sometimes worse.

The difference in approaches to, or styles in, mathematics that Weyl refers to, amplified by Noether's execration, is the difference between the explicit, sometimes laborious calculation of results and working on a higher conceptual plane marked by reasoning about the structure behind the problem. In the latter case, the mathematician sometimes proves things about the nature, for example, of the solutions to an equation (do they exist? are there a finite or infinite number of solutions?) without, perhaps, ever bothering to construct a single such solution.

Albert Einstein

The year 1905 is now called, by so many historians and popularizers of science, Einstein's "miracle year" (some prefer a fancy Latin version: *annus mirabilis*). In that one year, Albert Einstein, at the age of twenty-six, published five papers. Every one of the five papers was brilliant; some of them changed the course of human thought forever. One of those articles was to win him the Nobel Prize. It was a calculation that inaugurated quantum mechanics. From another of the 1905 papers, we get the one physics equation that everyone knows: $E = mc^2$ (although it doesn't appear in quite that form in the original). One of the five papers marked the birth of special relativity; its popularized ideas would haunt cocktail party conversations for decades and lead to innumerable sketches of trains on paper napkins. Another of the five papers would earn Einstein his PhD (in physics), two years before Emmy Noether earned hers in mathematics.

Einstein did all this while employed as a clerk in the Bern, Switzerland, patent office. His job was to examine patent applications. He liked it, because it was somewhat interesting, not especially demanding, carried a good salary, and left him plenty of time to think about physics. He joked that his desk drawer there, stuffed with theoretical calculations, was the "physics department." He didn't particularly like to teach classes.

Despite Einstein's contentment at the patent office, we have to wonder, what was he doing there? After graduating from university in 1900, he was the only physics student without a teaching assistantship. He spent a year unemployed, applying to various universities without success. Einstein's problem was very likely that he had simply offended his teachers. He cut many classes, preferring self-study. He took it upon himself to decide which classes were too boring or irrelevant to be worth his time. Teachers who are even minutely insecure tend to be easily put off by students who are too smart and who don't defer to their wisdom. And the young Einstein was not very diplomatic in general.

After a year of unemployment during which time he depended on his parents, who were not well off, for support, he finally got a job as a schoolteacher, first at a high school and later at another private school.[33] He enjoyed this work far more than he had expected, especially as both jobs left him enough free time and energy to work on physics problems. After a year of schoolteaching, he found a position at the patent office with the help of his friend and university companion Marcel Grossmann (as described later in the book, Grossmann gave him some even more crucial assistance). This patent work was more secure (the teaching jobs had been temporary) and was even more suited to him. In fact, his days at the patent office were some of his happiest.

Although Einstein produced much groundbreaking work in such a stunning variety of subjects while in Bern, the work that bears directly on our story is what we now call *special relativity*.

I will not give a detailed description of the content of this theory, as there are many excellent books and articles that provide that service, and it's not necessary for the thread of our story. However, we do need a basic overview, and, especially, we need to understand one particular way of looking at the theory. This angle on special relativity is not found in most elementary or popular accounts. It makes an intimate connection with invariant theory, the subject of Emmy Noether's PhD research. Noether was immersed in invariant theory when Einstein's paper came out.

First, why is Einstein's theory called a theory of "relativity"? The theory deals with the question of how you would describe things from, or relative to, different points of view. In this case, the points of view are different *reference frames*. That simply means environments that are moving at some constant velocity, which means at some constant speed in some particular, unchanging direction. If you're on a train moving smoothly along at a constant speed and I'm standing on the platform, we're in different reference frames of the type considered in this theory. Einstein's first relativity theory came to be called *special* in contrast to what came later. The general theory of relativity was, well, more general: it would include reference frames that are moving in *any* way.

The first explicit theory of relativity was laid out by Galileo, and we now call it *Galilean relativity*. It simply says that I on the platform will measure you as moving to the right (say) at whatever speed the train is going, and you'll measure me as moving to the left at that same speed. If you throw a ball in the direction of the train's engine, I'll see the speed of the train added to whatever velocity you imparted to the ball. Another example is the moving walkways now common in airports. Walking along at your usual pace, you look to the side and notice that the scenery is moving past more quickly perhaps than you're used to; the speed at which the scenery moves by is the platform speed added to your walking speed.

Galilean relativity is the instinctive theory of relativity that we believe in, usually without conscious awareness, unless and until we sign up for a physics class and have our instincts educated out of us. It's *obviously* true. It was the theory of relativity that persisted for over three hundred years, that Einstein inherited, and that he showed *could not be true* if some other things that we had learned in the meantime were true.

We will skip his arguments and proceed to some of the results. Briefly, however, the fundamental fact of life that Einstein used to prove that Galilean relativity needed to be replaced is this: in a vacuum, light has one speed, and that speed is *the same for everyone*, no matter the reference frame from which you measure the light. This means that if, as you're walking along the moving airport platform, you take out a flashlight, point it straight ahead, and switch it on, you'll measure (if you also had equipment for doing so) the light traveling away from you at this universal speed, represented by the constant c. So far, no surprise. But it also means that someone else, standing still on the nonmoving floor next to you, would measure precisely the same speed, c. This observation directly contradicts what we think we intuitively know about tossing balls on a train. (Of course the airport is not a vacuum, but its atmosphere affects the speed of light only minutely, and the idea is the same.)

Assuming this fact about the speed of light, which had been supported by experiment, and ruthlessly applying simple logic to ingenious thought experiments led Einstein to the conclusions of his special theory of relativity.

The constancy of the speed of light was also implied by James Clerk Maxwell's electromagnetic theory. In a sense, Maxwell's theory was the first unified theory in physics: the great Scottish physicist used the criteria of mathematical beauty and symmetry to blend the existing theories of electricity and magnetism into a set of equations that showed that each was an aspect of the other. These equations showed that vibrating fields of electricity and magnetism progressed through space in the form of waves—of light, heat, or radio waves—with a velocity that was a constant of nature and was independent of the motion of their source or of the observer. Maxwell's theory was an ever-present guide to Einstein while he created his own new physics; it was accepted by all, considered part of the ground truth, and it showed that Galilean relativity was not inviolate.

Among other things, these conclusions are that if you measure the flow of time, the duration of a second, on a reference frame moving relative to your own, you will measure it as being longer than a second in your own frame. In other words, if you stand on the platform and observe a clock on the train passing by, you'll see that clock ticking more slowly than the clocks on the platform with you. Time slows down on the train, compared with your own time. Why did no one notice this before Einstein? Of course, the effect, although real and now confirmed to tremendous accuracy by many experiments, is so tiny that you need either super-accurate clocks or speeds quite close to the speed of light to observe the effect. (And it's been confirmed both ways: by atomic clocks in airplanes and by observations that particles moving close to the speed of light "live" longer than they do when taking life at a more leisurely pace.)

Another result of the theory is that space itself is altered by speed. If you had the means to measure the length of the train car to extreme accuracy as it passes by, you would see that it's shorter than it was when the train was sitting still. The faster the train goes, the more contracted it becomes, in the direction of motion.

I won't say much more about these effects, except to dispel a common point of confusion, just in case: the people on the train, no matter how fast it goes, don't notice anything unusual about themselves or things happening *on the train*. Their clocks are slowing down as measured by the people on the platform, but the people are slowing down too. Time itself is slowing down, so there's nothing to notice. It's the same with space: there's no way for people to detect that things are shorter, because the rulers they would use to measure them are shorter as well. The length contraction is relative to *other* reference frames.

In 1902, Hermann Minkowski moved to Göttingen University. About the time that Emmy Noether was receiving her doctorate, he gave a lecture there about Einstein's recent special relativity theory. He had not only absorbed it thoroughly, but he had found a better (in his opinion) way to describe these transformations of space and time. It was, in fact, an elegant mathematical trick. "Einstein's presentation of his deep theory is mathematically awkward," he remarked. "I can say that because he got his mathematical education in Zürich from me."[34]

Yes, Minkowski was one of Einstein's math teachers at university.

Minkowski showed that in special relativity, the transformation of space and time between different reference frames is mathematically identical to a rotation of the space-time coordinate system—the set of axes on which we diagram the positions of objects in time and space (either one-, two-, or three-dimensional space). This observation made the unfamiliar familiar because, although Einstein's transformations were something new and strange in mechanics, everyone was already an expert in rotations. Whatever mathematical tricks and conveniences about rotations that we had learned from geometry could now make calculations in special relativity easier and more intuitive. In fact, we owe our concept of four-dimensional space-time to Minkowski. It combines the three space dimensions with the time dimension, but not in the way that Galileo or Newton might have done, where time and space retain their wholly separate identities. In Minkowski's space-time, the time and space coordinates are more intimate with each other: Minkowski's rotation mixed time intervals and space intervals together.

By showing that Einstein's space-time transformations were equivalent to a rotation, Minkowski had discovered a hidden symmetry in the equations of special relativity, one that Einstein hadn't seen. An important aspect of this mathematical point of view is to notice that as you rotate the coordinates, thereby changing the measurements of space and time, some things do not change. These important, unchanging quantities are the invariants of special relativity and are connected with the invariant theory in which Noether did her PhD work. As we'll see later, this import of a bit of invariant theory into relativity is the first precursor of Noether's achievement.

Minkowski's trick neatly brings out a radical consequence to relativity: space and time are not eternally separate. All you have to do is step on a moving platform, and space and time, formerly distinct concepts, become blended.

Minkowski noticed how revolutionary this consequence was: "Henceforth space by itself, and time by itself, are doomed to fade away into mere shadows, and only a kind of union of the two will preserve an independent reality."[35]

In his book about Einstein's creation of the general theory, John Gribbin suggests that the wide acceptance of Einstein's ideas, and even his academic success in general, owed a great deal to Minkowski's space-time formulation of the special theory.[36] While Einstein, as we've noted, came to appreciate Minkowski's work on relativity, it's doubtful that he shared this expansive view of its significance.

It took a while for Einstein to get the point of what Minkowski was up to. In fact, he made the crack, "Since the mathematicians have attacked the relativity theory, I myself no longer understand it anymore."[37]

He briefly considered Minkowski's four-dimensional space-time formulation a kind of pointless erudition and Minkowski's papers "needlessly complicated."[38] He put this in the context of his suspicion that the Göttingen mathematicians were sometimes more concerned with showing off than with trying to express things clearly, at least when it came to their physics-adjacent work.[39] Eventually, however, Einstein was to modify these opinions.[40] He not only came to appreciate Minkowski's approach but also eventually employed it in his own papers.

Emmy Noether, Math Teacher

Emmy Noether's father, whose health had always been precarious, grew progressively frailer after his daughter received her doctorate. Teaching was becoming more difficult.

Her PhD wouldn't get her an academic job. There were no such things as academic positions for women in Germany in the first decade of the twentieth century and most of the second. Of course, she knew this. She knew that Dr. Emmy Noether would not be able to be a professor of mathematics, while her male friends and colleagues with similar qualifications had found positions. She hadn't pursued her degree as a vocational qualification; she had done so out of pure hunger for knowledge and to gain research experience.

So she stayed in Erlangen and became her father's unpaid assistant. She also worked on her own research projects, which started out related to her thesis work: she pushed on for a while in invariant theory. She joined several mathematical societies and began to attend meetings. These associations let her discuss her work with a wider community and helped the wider community get to know her. She found the meetings delightful and exciting and enjoyed occasionally giving talks about her work. There was an active social scene in the evenings after the official sessions, and Emmy Noether was often there. She was almost always the only woman mathematician present—in other words, the only woman who was not there as the wife of a math man.[41]

Her thesis adviser, Paul Gordan, retired in 1910. His replacement's replacement, Ernst Fischer, became her most important mentor. For the next five years, they talked mathematics constantly. Noether created a voluminous set of notes of their discussions and sent him frequent postcards crammed with math, even though they both lived in Erlangen.[42]

Under Fischer's mentorship, Noether underwent a crucial transition as a mathematician. She turned away from Gordan's style of detailed, almost-algorithmic calculation and began to adopt a more abstract methodology. This was a style associated with Hilbert and with his methods of proof that Gordan had dismissed as theological. She embraced this new approach to mathematics enthusiastically and began to loathe the style in which she had initially been trained.

Her father became unable to teach or carry out his other duties. His daughter began teaching his courses as a permanent, unpaid substitute. She even supervised PhD students. She published additional papers, which were highly regarded, and continued to attend and speak at conferences.

In short, Noether's life at this moment resembled that of a top-flight mathematician at the beginning of a promising career, one destined to include the security and comforts of a prestigious appointment. She was teaching, supervising research, and, through her own research activities and lively participation in congresses and meetings of her colleagues, developing a rapidly growing international reputation. Her colleagues, the community of mathematicians, for the most part, accepted her as one of them—because she *was* one of them, one of the still-small coterie of insiders who spoke their recondite language.

However, Noether was not to enjoy the career of a top-flight mathematician, regardless of the esteem in which she might be held by her peers. German laws, regulations, traditions, and the dominant attitudes of society would ensure that, while the university would continue to benefit from her unpaid service and while she would continue to be allowed to enhance the reputation of German science, she would receive no official recognition for these things. Instead, she would watch a steady supply of less talented men pass her by on the career ladder.

This state of affairs would only begin to change after the end of the Great War.

While a few of her colleagues complained on her behalf and even battled to secure a position for her, her own attitude seemed to conform to Weyl's aforementioned description of her peace with her situation. There is no record of Noether ever uttering a word of protest about her treatment; nor is there any trace of bitterness or self-pity in any of her letters or recorded comments. On the contrary, she seemed to view her circumstances and the world in general through a humorous lens (but indeed not through rose-colored glasses). She didn't resent working without pay; she was in a position to help, so she did so. She expected nothing. She was gloriously happy to have any opportunity to pursue, teach, and talk about mathematics, the one thing that provided her with unalloyed joy.

Hilbert's fame continued to soar. He cemented Göttingen's status as the Western world's math headquarters. Now every aspiring mathematician wanted to study there, and Hilbert was the main reason. He was quickly becoming the world's first true math celebrity and turned the Göttingen math department into *the* mathematics destination that attracted students from all over Europe and the United States and as far away as Japan.

Hilbert's fame never seemed to have an untoward effect on him. A former student, Otto Blumenthal, noted that Hilbert accepted his acclaim with "a naive, mild pleasure, not letting himself be confused into false modesty."[43] By now, hundreds of people regularly filled the lecture halls at Göttingen to hear him talk, filling all the seats and sitting on the windowsills. Nevertheless, according to a someone who was there, "If the Emperor himself had come into the hall, Hilbert would not have changed." But not because of any sense of self-importance: "Hilbert would have been the same if he had had only one piece of bread."[44]

Although Hilbert's lectures were wildly popular, his unconventional approach sometimes got him into trouble.[45] Because he preferred to work things out on the blackboard rather than transcribe a carefully prepared lesson, he naturally got stuck now and then. If there was nobody around who could figure out how to proceed, he would just shrug and say, "Well, I should have been better prepared." Although this approach may sound like a recipe for chaos, it gave his students a glimpse of the struggles and creative effort that went into actually doing mathematics. In fact, it was this glimpse into the process of discovery that made his lectures so memorable for so many. The Hilbert classroom experience was, for most, a refreshing contrast to the perfectly organized and polished lectures of Felix Klein. Paul Ewald, later to become an important physicist and crystallographer but then employed as a "scribe" tasked with transcribing Hilbert's lectures, described the experience as witnessing Hilbert seemingly creating new mathematics on the spot as needed rather than regurgitating well-known results.[46]

Early in the 1900s, Hilbert was awarded the German version of, roughly, a British knighthood.[47] However, he was impatient, even abusive, toward those who insisted on addressing him using the title.[48] Hilbert didn't much care exactly how he was addressed, but excessive formality and, especially, obsequiousness were intolerable to him.

Klein and Hilbert were an interesting study in contrasts. Klein had also received the knighthood and preferred that he be addressed with the title that went with it. The American mathematician Norbert Wiener went to visit Klein after the German professor had retired. Wiener knocked on his door and asked the housekeeper if the "Professor" were in. She corrected him sternly, saying that Klein was indeed at home, but she substituted his exalted title in place of "Professor."[49] When a former student was asked how Hilbert preferred to be addressed during those years, he replied, "Hilbert? He didn't care. He was a king. He was Hilbert."

During Hilbert's tenure at Göttingen, he made a profound impression on a long series of students and colleagues, many of whom were to become renowned for their own contributions to mathematics and science. Max von Laue, the future physics Nobelist but then a student attending some of Hilbert's lectures at Göttingen, said of his professor, "This man lives in my memory as perhaps the greatest genius I ever laid eyes on."[50] In contemplating the significance of this remembrance, we should keep in mind the collection of first-rate scientific minds that von Laue had to have rubbed shoulders with over his lifetime.

The mathematician Harald Bohr remembered Hilbert this way: "Over the whole life of Göttingen shone the brilliant genius of David Hilbert, as if binding us all together. . . . Almost every word he said, about problems in our science and about things in general, seemed to us strangely fresh and enriching."[51]

Minkowski later wrote to Hilbert that "one can learn much from you, not only in mathematics, but also in the art of enjoying life sensibly like a philosopher."[52]

It would be a mistake to think of Hilbert as a distracted, absent-minded-professor type, despite his constant preoccupation with the most rarefied realms of pure mathematics; he was not without shrewdness in human affairs when the occasion called for it. Such an occasion arose when his rocketing fame led to an offer of a prestigious professorship at Berlin.[53] His students were afraid that such an opportunity was too tempting to refuse, but they tried nevertheless to prevail on Hilbert to stay at Göttingen. Hilbert seemed preoccupied and, to their dismay, did not allay their fears. They knew nothing, however, of the machinations that their beloved professor was setting in motion behind the scenes, scheming to use his offer of a position as leverage. With diplomatic adroitness, he managed to pressure Klein's accomplice, Friedrich Althoff, to create a new position at Göttingen and allow it to be filled with Hilbert's old friend, Minkowski. When the dust had settled, everyone at Göttingen was overjoyed, including Hilbert himself; his scholarly life would now be further enriched by the presence of one of his favorite mathematical companions. Minkowski would turn out to have an important influence on Hilbert, convincing him, critically, to delve further into physics.[54]

Göttingen is more than just a place of employment, far more than just a school. The old university is one of the central characters in this story: the spirit of the place itself, the weight of its splendid history, its embrace of the intellectual journeys that unfolded both within its stone walls and along paths through its forests, elevate it to the status of an active participant in the momentous discoveries of Hilbert and his friends. American physicists Leon Lederman and Christopher Hill describe the German university of this era:

> The university of late nineteenth- to early twentieth-century Germany . . . was a profoundly influential community, particularly in the sciences and mathematics, where it enjoyed the reputation of being the best in the world. It was a place of the highest academic standards of the age, the birthplace of quantum mechanics and Einstein's general theory of relativity as well as most of modern mathematics. . . . Here ethnic minorities found a tolerant, open, and receptive community, a place to flourish that offered a respite from an outside society of staunch national conservatism. It was a quiet, meditative environment, a community of scholars with a common deep and abiding love of their abstract pursuits.[55]

Lederman and Hill speak not of Göttingen in particular but of German universities generally; Göttingen happened to be the most highly regarded at the time in mathematics and the sciences. Of course, this toleration, freedom, and meritocracy had limitations: ethnic minorities, yes; women, certainly not. There was another limitation, relating to class. This restriction was not enforced by the universities themselves directly but was a de facto constraint arising from the German educational system. Only those with a certain degree of means could prepare themselves to pass the examination that would allow them to partake of this meritocracy.

How much did the intimidating position of German scholarship in these fields owe to their formidable system of universities rather than to, say, an accidental agglomeration of brilliant minds or to some other cultural or historical circumstance? It's surely impossible to give any kind of quantitative answer to questions such as these, but just as surely, the influence of the university was profound.

While his renown and the scope of his responsibilities continued to grow, Hilbert remained utterly uncompromising in his principles. He was aghast at Germany's annexing of territory and had refused to sign, at the outbreak of World War I, the notorious "Manifesto to the Civilized World" defending Germany's territorial aggression.[56] As an indication of the nature of this document, which was signed by ninety-three reputable German thought leaders, it warned against the "Russian hordes allied with Mongols and Negroes unleashed against the White race."[57] Nearly all his colleagues (even Klein and Max Planck) and other prominent people all over Germany had felt obliged to put their names to this screed (which, in later years, caused some of them shame and others to profess their lack of full comprehension of the extent of Germany's crimes. As Upton Sinclair famously said, "It is difficult to get a man to understand something when his salary depends on his not understanding it"). As a result, his patriotism was suspect: no small matter during wartime.

Meanwhile, as the war got underway, Hilbert continued to outrage the older faculty by ignoring their ideas about how someone in his position should behave.

After he moved to Göttingen, Hilbert's research interests moved away from invariant theory to such topics as logic and the foundations of mathematics. His work was marked by an ever-increasing formalism and a focus on the structure of mathematical systems.

Hilbert pioneered a highly abstract, axiomatic approach to mathematics. The meaning of this approach may not be very clear to those who find mathematics so inherently abstract that additional distinctions seem meaningless. But there are different styles of mathematical research. In Hilbert's day, confidence resided in concrete constructions of results and solutions, not merely in invariant theory but in general. Hilbert aimed to avoid detailed calculations in favor of demonstrating, sometimes indirectly, the logical necessity of the results he was trying to prove. Eventually, his critics were forced to concede the usefulness of Hilbert's methods, and the axiomatic flavor of mathematical research has become part of the mainstream.

Sharon McGrayne, in her book *Nobel Prize Women in Science*, places Hilbert's predilection for abstraction in a broader cultural context: "Early twentieth-century intellectuals—including mathematicians, artists, architects, musicians, dancers, writers, and physicists—were fascinated by the concept of abstraction. Eager to strip reality of its special, individual peculiarities, they sought general principles that always hold true. David Hilbert, considered the greatest mathematician since Carl Friedrich Gauss, was using highly abstract methods at Göttingen. In Erlangen, Noether began applying his approach to algebra."[58]

McGrayne also points out that Emmy Noether began her mathematics career somewhat in the shadow of her well-known father and was known as Max's daughter but that, eventually, Max became identified as Emmy Noether's father and is so identified today.

Peter Freund, in *A Passion for Discovery*, adopts a similar point of view. He notices the similarities between the increasing abstraction of early twentieth-century mathematics and physics, and the emergence of a higher degree of abstraction in the arts, which he feels is represented in the works of Wassily Kandinsky, Arnold Schoenberg, James Joyce, and others.[59]

Hilbert's concern with the logical structure of mathematics found one of its early expressions in *The Foundations of Geometry*, a book he published in 1899.[60] This book, and the approach it illuminated, was to have a deep impact on the subsequent development of mathematics.[61] A year after its publication, in his famous speech wherein he laid out his famous twenty-three problems, problem number six challenged scientists to apply the axiomatic approach used in his geometry book to the various fields of physics.

Hilbert may or may not have remembered Emmy Noether's semester in his classes in 1903. It's likely that he did, if for no other reason than the presence of a female student would have been unusual though, at Göttingen, not unique. He had probably encountered a younger Emmy at least once, when visiting with her father. In any case, by 1913, he would have been aware of her reputation and activities as a mathematician. The world of mathematics was still quite small. An active mathematician would generally be aware of the other productive members of the discipline, especially, as was the case with Hilbert and Noether, when they were conducting research in the same or adjacent areas.

In 1913 Emmy and Max Noether visited Göttingen to work with Hilbert and Klein in composing Paul Gordan's obituary. The "King of Invariants" had died.

We do not know how or when Hilbert first became aware of Emmy Noether; but even if her auditing of classes shortly after her PhD went unnoticed, she left a strong and favorable impression on both Hilbert and Klein from her 1913 visit.

Noether's series of papers that began to appear after her PhD could not have failed to attract the notice of David Hilbert.[62] He would have been even more likely to have noticed her papers in those days, before the proliferation of publications grew out of control. Noether published papers extending Hilbert's basis theorem and attacking related problems, proved a conjecture made by Hilbert in 1914, and partly solved Hilbert's fourteenth problem, from his famous set of problems for the new century. She also established herself as an expert in something called the *theory of differential invariants*.

Whether or not Hilbert remembered Noether personally from previous encounters, the woman whom he now saw had undergone a transformation. Her appearance was no longer at all that of a conventional middle-class woman. She had gone full math mode.

She had pared her clothing down to the practical minimum and adopted an unusually short hairstyle (to become more fashionable some decades later), creating a spectacle of herself of which she seemed unaware. She had for some time begun to wear what others described as strange clothing for a proper young woman, such as plain black frocks that caused one student who encountered her to remember that she looked to him like a railway conductor. Altogether, she was such an odd sight that people sometimes stopped and stared at her on the street.

By the end of the obituary-writing session, Hilbert and Klein had invited Noether to come to Göttingen to engage in mathematics research alongside them.

What Is Important Is Gravity

At nearly the same time that Emmy Noether accepted the Göttingen invitation, Einstein accepted an invitation to come to Germany as a research professor at Berlin. One of the main attractions of the job was that he wouldn't be required to teach.

Einstein had turned physics upside down with his 1905 papers, but he was just getting started. Immediately thereafter, he began his gargantuan project to turn the special theory of relativity into a general theory.

The special theory was limited to considering what physicists call *inertial reference frames*, the above-mentioned frames of reference moving at constant velocity. The *general* theory would relax that condition and include any reference frame, in particular frames that were accelerating. To a physicist, *accelerating frames* means frames changing speed or direction, or both.

The general theory was to be a theory of gravity. Of course, we already had one of those: Newton's law of gravity, which describes a force between any two objects as proportional to the product of their masses and decreasing as the square of the distance between them. It had served humankind well for centuries. Most impressively, it had allowed Newton to explain, with one simple law (plus his laws of motion), where Johannes Kepler's elliptical orbits came from—an explanation that amounted to the final consummation of the Copernican revolution.

Copernicus's revolution replaced the geocentric universe with the sun-centered one. Before Copernicus, the celestial bodies had all revolved around us, in large circles with smaller circles imposed upon them: the notorious epicycles. After Copernicus, we and the other planets revolved around the sun, but the epicycles were still there, needed to make the model agree with observation. They were needed because, unknown to Copernicus, the orbits were not circles—something unthinkable to him, a member of the ancient cult of circular perfection.

Kepler replaced the circles with ellipses, along with several rules for the varying speed of orbit along each ellipse. This model replaced the complex system of epicycles with a few simple rules that agreed with observations, but it was a purely geometrical solution: nobody knew where these ellipses came from.

Newton showed that his simple rule for gravity was the cause of Kepler's ellipses, replacing an arbitrary geometrical description with a true law of nature, a causal system. It was also a great unification: the force that holds the universe together was the same one that caused the apple to fall from the tree. This unification explains the importance of Newton's universal law of gravitation and why it was a turning point in our understanding of the universe and our place in it.

There were some problems with Newton's gravity, however. Some of them were conceptual, and one of them was empirical. The empirical problem was a long-known anomaly seen in Mercury's orbit. Its ellipse was not eternally unchanging, but *precessed* by a tiny amount every year. Noncircular ellipses don't have the perfect symmetry of circles; they're elongated, so they have two radii, or axes, of different lengths. The precession of Mercury's orbit means that (either one of) these axes slowly rotates, rather than pointing in the same direction forever. The orbit's direction in which it's flattened slowly changes. Another way of saying this is that the planet, after completing one of its years, doesn't return to precisely the same place: its orbit is not closed.

Now, none of the orbits of the planets in our solar system is a perfect, closed ellipse. That's because each orbit is perturbed by the presence of all the other orbiting planets, especially by ones nearby. When we say that the orbit of Mercury exhibits an anomaly, we mean that there is an extra observed precession beyond what can be explained by the interference of the known planets. For a long time, astronomers had assumed that this orbital irregularity must be due to a planet that we had not yet observed directly. However, when they looked for it at the place where calculations suggested it should be found, they saw nothing.[63]

Einstein was one of the small number of physicists who thought that it was the law of gravity itself that needed to be modified. We shouldn't be looking for an unknown planet; we should be looking for new physics. He was the only one who succeeded in finding that modification.

The conceptual problems with Newton's law of gravity were several. One was the magical action-at-a-distance character of the law. According to Newton's model, the force of gravity instantaneously reached through expanses of space without limits of any kind. As the masses that caused the force moved, the forces, even on other masses millions of miles away, changed direction instantly. If the sun were to suddenly vanish, its gravitational influence on our planet would vanish without delay. And if another star were to pop into existence, we would experience its tug without having to wait for its influence to travel through space. Einstein was not the only scientist bothered by this model of reality, but he was the only one who knew what to do about it.

The other major head-scratcher was the mysterious fact that the masses in two of Newton's laws seemed to be the same, for no good reason that anyone had ever found. The force of gravity that pulls you down to the ground is proportional to your mass. Newton's second law of motion says that the force with which someone has to push you to change your velocity is also proportional to your mass. The two laws have nothing to do with each other, but the same mass appears in both. Measurements showed that we were, in fact, dealing with the same mass. The gravitational mass was the same as the inertial mass.

One of the consequences of the identity of the two masses is the well-known observation made, famously but perhaps apocryphally, by Galileo in his Tower of Pisa experiment: heavy and light objects undergo the same gravitational acceleration near the earth's surface. Air resistance complicates the demonstration, but the experiment is routinely repeated in the classroom with an evacuated tube containing a feather and a rock. The demonstration was also repeated, with a feather and a hammer, on the moon by an Apollo 15 astronaut.[64] Our expectations, primed by our experience within the atmosphere, are thwarted when we see the rock and the feather falling side by side, and the effect is delightful and memorable. The force of gravity is stronger on the rock (it weighs more on the scale, because it is pulled down by a stronger force). However, it resists acceleration to the same degree; the effects balance each other exactly, leading to the gravitational acceleration near the ground being considered a constant property of the earth.

This was one of the fundamental mysteries in classical physics, and yet no one was working on it, because there seemed to be nowhere to start.

It was these conceptual problems with Newton's gravity that motivated Einstein to think about the possibility of a better theory and to remove the artificial limitation to inertial reference frames in his special theory. This generalization to a wider class of reference frames would ultimately appear as the replacement of the rotational invariance described earlier in this chapter with more general classes of transformations, with the uncovering of a hidden symmetry that explained gravity and made the identity of the two forms of mass inevitable.

Einstein's puzzling over the apparent identity of the two forms of mass is related to what we now call the *principle of equivalence*. This principle can be formulated in several ways. One way is to insist that the two types of mass are identical because of a fundamental symmetry in nature—that the laws of physics must take the same form whether you are in a gravitational field or in a region of space with no gravity but, say, in a spaceship undergoing an equivalent acceleration.

Einstein proposed one of his famous thought experiments to explain the principle. Suppose you are in a sealed box and conduct experiments and make observations, as one does, and that your observations are consistent with you and the box sitting at rest on the surface of the earth. When you drop things, they accelerate downward at 9.8 m/s^2, meaning their speed increases by 9.8 meters per second every second. Pendulums behave in their customary way, and when you step on a scale, you see you have not made any recent progress with your diet.

Einstein pointed out that all these observations would be unchanged if, instead of sitting on the earth, the box were being towed through empty space with an acceleration of 9.8 m/s^2. Aside from tidal effects (avoided by confining measurements to a single point, avoiding changes in gravity with height), there is no way, from inside the box, to tell which situation you are in. The scale conveys the same disappointing news, because your "weight" is reproduced exactly by the apparent force created by the box's acceleration.

Since it is impossible in principle to distinguish between the two situations, they must be fundamentally the same, and their sameness should be reflected in the equations describing gravity and motion. The identity of the gravitational and inertial masses would cease to be a mystery if inertia and gravity themselves were understood to be somehow identical.

Carrying the principle of equivalence to its logical conclusions eventually led to the equations of general relativity, the theory considered by many, including the great theoretical physicist Lev Landau, to be "probably the most beautiful of all existing physical theories."[65] A recent *Economist* article muses on the theory's appeal: "The theory explained, to begin with, remarkably little, and unlike quantum theory, the only comparable revolution in 20th-century physics, it offered no insights into the issues that physicists of the time cared about most. Yet it was quickly and widely accepted, not least thanks to the sheer beauty of its mathematical expression; a hundred years on, no discussion of the role of aesthetics in scientific theory seems complete without its inclusion."[66]

Einstein published an early version of the theory in 1907, where he attempted to modify Newtonian gravity to take into account the finite signal speed required by special relativity, where instantaneous action at a distance was not allowed. This paper represented the first time the principle of equivalence was explained, the principle that Einstein called his "happiest [or luckiest] thought."[67]

Einstein's fame among physicists grew. Professors from around the world started to visit him in his Swiss patent examiner's office. Many were surprised to find that the name attached to so many important papers belonged not to an established professor of physics at an eminent university but to an informal young man sitting at a modest government desk.

Of course, Einstein liked it there. Some of his friends, however, thought it would be better if he had a more normal job for a serious scientist. He eventually agreed to take a professorship at the University of Zürich but only after the university agreed to raise its salary offer to match what the patent office was paying him.[68] His ambivalent attitude to his new academic status may be gleaned from what he wrote to a friend: "So now I also am an official member of the Guild of Whores etc."[69] It was 1909, two years after Emmy Noether's PhD.

From there he moved to a professorship at Prague, then back to Switzerland, and then, in the spring of 1914, to Berlin. Einstein, Hilbert, and Noether were now working within a circle with a radius of about two hundred miles. In less than a year, Noether would move to Göttingen, and the circle would contract.

Emmy Noether at Göttingen

Noether had lingering responsibilities at Erlangen. Rather, her father did, and she had taken them on. But by April 1915, she found herself able to make the move to Göttingen. She began to work alongside Felix Klein, David Hilbert, and others at the renowned mathematics department, unofficially, without pay. Women were still not allowed on faculties. World War I was being fought far from the idyllic campus, but its influence was felt. In a time of nationalistic entrenchment and redoubling of conservative attitudes, the old rules were not about to be changed.

Two weeks after moving to Göttingen, Noether received word from Erlangen that her mother had died.[70] The next months were filled with chaos, as she traveled back and forth between Göttingen and her hometown. Her father, by this point, was quite frail, and her mother had been his chief caretaker. Despite the frequent trips that the situation demanded from Emmy Noether over the next year or so, she managed to get some important work done.

Much of this work had to do with physics. Hilbert, for some time, had been preoccupied with various questions in theoretical physics. Klein, as well, was deeply interested in the field. Emmy Noether was not a physicist and had entertained no serious interest in any scientific research, apart from pure mathematics. However, Hilbert, with remarkable prescience, had realized that her particular knowledge and skills stood an excellent chance of bearing directly on a problem of acute interest to him.

This was the problem of general relativity that Einstein was working on. Hilbert had followed this work with fascination and had made several attempts, starting in 1912, to get Einstein to come to Göttingen and discuss general relativity with him and the other professors in the math department. So far, Einstein had declined. (Purists insist that the theory should be called the *general theory of relativity* and never *general relativity*, but I will interchange these descriptions with abandon. Most physicists certainly do so in practice.)

Hilbert recognized that questions of symmetry and invariance lay at the heart of general relativity. He knew that Noether was a leading expert in invariant theory and related areas of mathematics that would bear directly on the increasingly arcane machinery that seemed to be required for this new theory of gravity. After bringing her to Göttingen, he knew that, along with himself and Klein, he now had in one place the three mathematicians in the world with the greatest chance of making headway on this strange and interesting problem.

Einstein Rides the Train

In July 1915, a young, still handsome, but scientifically frustrated Albert Einstein boarded a train from Berlin, headed for Göttingen. He had finally accepted an invitation from Hilbert. The physicist had never met the mathematician, who was certainly persistent. In Einstein's notes to himself, he discounted the likelihood that anything particularly worthwhile would come of the visit. But he knew of the reputation of the place and knew some of the people there, and he had decided that he might as well give it a shot. The previous eight years had been exhausting drudgery for Einstein, as he tried to mathematically express the ideas in his new theory: a theory of how matter and energy created gravity by changing the shape of space and time—a theory that described the structure of the universe. He had come up with a system of ten equations that came close to succeeding. But despite years of intense effort to get those ten equations to dance together with the intricate steps required, they would not cooperate.

The choreographer was Einstein himself and his philosophically motivated requirement that his new theory be supremely objective.

The term *objective*, in this context, has a specific and technical meaning. Briefly, the idea was that the math that describes space-time and its interaction with mass and energy should provide the same description from the point of view of any reference frame and any coordinate system. The equations should reflect an objective reality in a way that was independent of both how time and space were measured and the point of view of the observer. He had met this requirement for the highly restricted case of inertial reference frames, but that was no longer enough.

This was the source of Einstein's frustration. He had no expectation that the mathematicians waiting for him at the other end of his train journey would solve this problem for him. But he permitted himself a glimmer of hope that they might at least understand what he was trying to do.

The circle had contracted to a singular point.

2

Gravity

"I have hardly come to know the wretchedness of mankind better than as a result of this theory."

Albert Einstein, David Hilbert, and Emmy Noether were now all present at Göttingen at the same time. It was gravity that had pulled them together. Specifically, it was Einstein's relentless and revolutionary project to supplant the theory of gravity, which had ruled the universe since 1686, with something better. To do this, he had been obligated to stretch his mathematical chops to their limit, straining to speak a language that he had to learn through immersion. It was fortunate for him that he had finally accepted Hilbert's repeated invitations to visit the world's mathematical epicenter. For here he would encounter not merely a receptive and encouraging audience but also some of the most talented mathematicians in the world. These colleagues would help carry him through the following four months to the completion of his project.

The Göttingen denizens each brought a unique, individual perspective and set of talents to the task. Hilbert could speak the relevant mathematical language creatively and bend it to his own ends, but some of its idioms were still unfamiliar to him. Noether spoke the language like a native and would become the translator and guide for the rest of the group. She alone was able to penetrate into its deepest secrets. Felix Klein was of the previous generation but keenly interested in physics and still well able to absorb new ideas in mathematics. He would often assume the role of liaison, relaying Noether's explanations and results to Einstein in a stream of correspondence that began immediately after the physicist's visit and lasted until his publication of his theory of general relativity, and beyond. (Noether and Hilbert also corresponded directly with Einstein during this same interval, but much progress was made during discussions among the Göttingen group, and Klein would at times write to Einstein about the outcomes.)

Within this crucible would be forged Noether's great discovery, the central idea in this book, the theorem described in detail in Chapter 3. To understand where the theorem came from and why it exists, we need to understand the circumstances that gave it life: these were connected to the need to comprehend gravity and the structure of the cosmos.

That's one purpose of the current chapter, but there is another. Because her name is on the paper announcing Noether's theorem, history acknowledges who discovered it. But Noether's fingerprints are also on general relativity itself. As I argue here and elsewhere, if Einstein had not had extensive assistance from several others, he could not have gotten general relativity to work, and therefore it's not unreasonable to describe the theory as having several authors.[1] This argument does not diminish Einstein's critical role in applying his supreme gifts of physical intuition and in refusing to give up. Sustained by a conviction born of that intuition, he persisted over many years, certain that he was on the right track. Several historians have resurrected the contributions of Marcel Grossmann and Hilbert, at least, to the formulation of general relativity. Regrettably, however, Noether is rarely mentioned in this connection; I wish to correct this omission.

There are two ways in which Noether should be considered among the (secondary) authors of general relativity, alongside Grossmann,

Hilbert, and perhaps one or two others. First, correspondence during this period and later comments by Einstein and Klein demonstrate that she gave repeated and focused help to Einstein to help him understand his equations and to guide him to his result. Second, Hilbert's own, mathematically quite different, formulation of general relativity, for which history grants him credit, also depended on Noether's work. We'll return to this subject in the next chapter. For now, it's sufficient to appreciate that both approaches to gravitation, Einstein's and Hilbert's, benefited from Noether's collaboration and would quite likely not have succeeded without her.

Her critical contributions to general relativity became one of several ways in which Emmy Noether helped invent modern physics.

Hilbert and Physics

By 1915, Hilbert sat like a spider at the center of an academic web, aware of every vibration in the world of mathematics and physics, able to summon the likes of Einstein, who had already become renowned among physicists throughout the world, to Göttingen to satisfy his own curiosity.

Hilbert's background and research up to this point had been in pure mathematics. Why the sudden fascination with physics?

Hilbert was no dabbler. His fingerprints are all over the machinery of theoretical physics. This was partly due to his ecumenical stance toward mathematics and science—to the invisibility, for him, of the boundaries that more conventional scholars believed separated the various disciplines. His imprint also arose out of the depth, breadth, and penetrating originality of his investigations, which provided physicists with new techniques to formulate and solve problems. He even, as in the case of something now called *Hilbert space*, invented entire mathematical structures and environments within which physical scientists could arrange their calculational furniture.

In fact, Hilbert maintained a lively interest in theoretical physics for most of his career. He even arranged to have a physicist assistant keep him apprised of important developments in the field and tutor him on whatever physics he couldn't figure out for himself. Some of these physics scribes went on to become distinguished physicists in their own right. One of them was Max Born, later to become one of the founders of quantum mechanics and a Nobel Prize winner. Born fondly remembered his times with Hilbert and Hermann Minkowski: "It was a wonderful learning time for me. Not only in science, but also in the things of human life."[2]

Leo Corry is a prominent historian of mathematics and physics at Tel Aviv University and the president of the Open University of Israel.[3] In a penetrating essay from 1999, he shows that the view of Hilbert as a mathematician who developed an interest in the physical sciences would miss the point.[4] In Hilbert's mind, the worlds of physics and mathematics were not clearly separated in the way they have become in recent decades. The walls between these disciplines that we take for granted today are partly the result of the artificial boundaries that define the map of the modern university, with its jealously guarded little kingdoms. But they are also a by-product of the increasing specialization that's an irreversible feature of all the sciences. In the early decades of the twentieth century, a physicist could read and actually understand, to a good degree, all the theoretical research papers appearing in the journals every month. Similarly, a mathematician, and certainly one with Hilbert's eclectic interests, would be at least conversant with every active area of mathematical research and aware of all recent developments. For this reason, it shouldn't be surprising that Hilbert had remained aware of Noether's research activities and their potential relevance to his own ongoing interests, including those touching on physics.

Today, the research journals have proliferated beyond the imagination of a late nineteenth- or early twentieth-century physicist or mathematician, having undergone uninterrupted exponential growth since

World War II.[5] Researchers can no longer understand papers in specialties in which they are not trained—even those closely related to their particular fields.

But let us go back to 1905 or 1915 and try to see things from Hilbert's point of view. The sciences and mathematics were less fragmented, and their separate jargons had not yet become mutually impenetrable. Hilbert was unusually wide-ranging in his interests and abilities, and he seemed able to maintain a mental landscape of practically all of mathematics and a good deal of physics. For Hilbert, however, this attitude was not a matter of nurturing an awareness of separate disciplines. Hilbert considered mathematics and physics to be, if not largely the same enterprise, then close cousins under the umbrella of science, even if the current state of development of a particular theory in physics did not make this association obvious.

The most instructive example of this viewpoint, as Corry points out, is Hilbert's attitude toward geometry.[6] He thought of geometry as an empirical science whose axioms should reflect our experience of physical space. Its theorems could be tested by performing experiments, such as by actually measuring the angles of a triangle to confirm that they added up to two right angles. The correctness of geometry is demonstrated daily by our ability to navigate, survey, and build with its postulates and results as our guide. The discovery of theorems implied by the axioms and definitions followed the rules of logic and mathematics, just as in a "pure" mathematical field (if there were such a thing), but those axioms and definitions were reflections of our physical experience. Euclidean geometry, and Hilbert's reworking of it, belonged to physics as well as mathematics.[7]

We can thus understand Hilbert's concern with the logical hygiene of a mathematical theory such as geometry as partly motivated by his desire to create the best possible science of the physical world. When the axiomatic structure of a theory is clarified, in Hilbert's view, all that really remains is structure itself. In this way, as in several others,

Hilbert was ahead of his time. Applying these notions to geometry, he is often quoted, to this effect, as claiming that "one must be able to say at all times—instead of points, straight lines, and planes—tables, chairs, and beer mugs," although, sadly, there is no authoritative source for this highly suggestive and lovely remark.

The meaning of the reference to beer mugs and so forth is that the *structure* of an area of mathematics, rather than how we think of the objects it refers to, defines its essence. An analogy to games might make this more transparent: if we decided, when playing chess, to call the pawns "cars" and the knights "donkeys" but kept all the rules the same, we would still be playing chess. The names of the pieces are incidental. Some mathematicians were uncomfortable with Hilbert's "formal" approach to mathematics, just as I'm uncomfortable trying to play chess using an excessively artistic set with abstract sculptures for the pieces—but that's partly because I'm a bad chess player who doesn't think at a high-enough level and who needs the comfort of familiarity: knights that look like horses. (For a brief reminder of the logical structure of mathematics, see the section "Axioms, Definitions, and Theorems" later in this chapter.)

This unity of thought may help explain why Hilbert's results are all over the applied-mathematics toolshed that any physicist enters to get things done. With the notable exception of Hilbert space, these various tools and theorems often exist without his name attached.

Hilbert considered the compartmentalization of the sciences an artificial one; he had written in his notes that "the separation of the sciences into professions and faculties is an anthropological one, and it is thus foreign to reality as such. For a natural phenomenon does not ask about itself whether it is the business of a physicist or of a mathematician."[8]

As mentioned earlier, Noether's mathematical style and interests began to resemble Hilbert's after she decisively turned away from the earlier style represented by her PhD thesis. Her concern with abstraction and structure grew continuously from that moment in her career and, in the opinion of some chroniclers, eventually surpassed that of Hilbert himself in its exclusive residence at the most formal levels of mathematical thought.

But the preceding paragraphs lead us to consider one difference in Noether's and Hilbert's interests and obsessions. Noether was always to remain exclusively immersed in pure mathematics. She would be utterly indifferent to, and perhaps even impatient with, any talk of possible applications of her research.

Hilbert, in contrast, was interested in doing for physics what he had done for geometry. His *Foundations of Geometry* is largely devoted to teasing out which theorems depend on which axioms and, just as importantly, which axioms are superfluous for the derivation of certain theorems. He was interested in proving both the consistency, completeness, and mutual independence of sets of axioms, thereby making it crystal clear which parts of our knowledge depend on which assumptions. This work was to have a profound influence on the subsequent development and style of mathematics. However, it belongs to the era before mathematician Kurt Gödel's world-shattering results in logic and, for that matter, just barely before Bertrand Russell's enunciation of his famous paradox. So, in a sense, Hilbert's work belongs to a logically naive era of mathematics.

Russell showed that a version of the ancient Greek liar's paradox ("this statement is a lie") persisted into what we now call *naive set theory*. His example asks us to consider a town where the barber shaves all the men who do not shave themselves, and only those men. He then poses the question "Who shaves the barber?" If the barber shaves himself, then he doesn't shave himself, because he shaves only those men who don't self-shave. But if he doesn't shave himself, then he does, because he shaves all those who don't shave themselves. Thus Russell showed that innocent-looking definitions of sets (of barbers or anything else) can lead to paradoxes.

Gödel's results proved that most mathematical systems contained true statements that would elude proof within the system, exposing the limits of axiomatization. This work forced mathematicians to accept that while attempts at comprehensive formalization, such as Hilbert's, led to valuable insights, they would never be able to contain all of mathematics.

Axioms, Definitions, and Theorems

I've been using terms related to formalization, axiomatics, and abstraction rather freely in the foregoing; at this point it would be worthwhile to pin down more carefully what these ideas entail. This will help to clarify the flavor of just what it was Hilbert, Noether, and like-minded mathematicians were working on.

To this end, I offer a brief digressive refresher for those who may be a bit cloudy on the relationships among definitions, axioms, and theorems, and what these terms mean.

Mathematicians are concerned with deriving true statements from other things that they know or assume to be true. They need to be on solid ground. They need to know what the statements mean, exactly, and to be sure that their proofs are valid.

Any bit of mathematics begins with *definitions* of its basic terms, especially any new terms introduced. Modern mathematics is characterized by extremely careful, detailed definitions. The discipline has been burned in the past by sloppy or insufficiently rigorous definitions, and mathematicians have learned to get everything right up front. You don't see this style of definition in high school mathematics, but it's all over math at the advanced level.

One place you see formal definitions is in Euclidean geometry, which most people have a course on in high school. You see definitions of a point, a line, and so on, typically following the ancient versions pretty closely. (But often with less charm. Euclid defined a *point* as "that which has no part" and a *line* as "a length without breadth." There is some austere poetry in these pronouncements.) Many students wonder why they're forced to sit through tedious statements of things that are, after all, intuitively obvious. They don't know how lucky they are: modern definitions tend to be far more annoying in their extreme specificity. This obsession with detail, mathematicians have learned, helps in avoiding mistakes and ambiguities later, when the definitions are used in proving theorems.

Next come the *axioms*. These are statements that we assume are true. They usually seem as if they *must* be true; we can't imagine a world in which they're false. If they're so obvious, why can't we prove them? Why do we need to assume?

We have to start somewhere. Every mathematical system rests, logically, on its definitions and its axioms. Without assuming, absent proof, that some things were true, there would be nothing on which to base the subsequent statements that we want to prove. In a correctly constructed mathematical system, the axioms are all independent of each other; no axiom can be deduced from the others. If it could, it wouldn't be an axiom, but a theorem.

Euclid wrote down five axioms for his geometry. The first one was this: "Given two points, there is a straight line that joins them."

(Some writers claim that Euclid laid out ten axioms in the *Elements*, the book that contains his geometry, and they're not wrong. But Euclid called the second group of five axioms not postulates but "common notions." The first five are about his geometrical things: points, lines, and so forth. The second five, the common notions, are about quantity. They contain such statements as "Things equal to the same thing are also equal to one another.") Euclid's fifth axiom is, in his original version, oddly longer and more detailed than the others. It talks about two lines and a third line intersecting them. It then says that you can tell if the original two lines will ever meet by looking at the angles made between them and the third line and comparing their sum to the sum of two right angles. The other axioms are no-brainers: clearly true. This one requires some thought or sketching on paper before it becomes clear that Euclid is talking about parallel lines and that this axiom, too, is obviously true.

Another word for axiom is *postulate*. The statement about parallel lines is famous and usually referred to as Euclid's fifth postulate or the *parallel postulate*.

I'm going on at some length about axioms to clarify some of the things that first Hilbert and then Emmy Noether were examining.

Shortly after arriving at Göttingen, as mentioned earlier, Hilbert published a thin book called *The Foundations of Geometry*, in which he reworks Euclid's legacy. Part of what Hilbert did in this book was to write down better versions of a couple of Euclid's axioms.

"Better" in this context means that he found simpler, clearer statements that had the same content and from which the same theorems could be derived. He didn't *change* Euclid's geometry; he said it better. That Hilbert could improve on a cornerstone of classical mathematics that had been etched in stone for two thousand years tells you something about his intellectual stature.

Hilbert judged the first axiom imprecise and replaced it with an improved version. He also constructed a more useful version of the much-discussed fifth postulate and cleaned up the structure of Euclid's edifice throughout. In fact, the versions of the axioms usually taught in school are closer to Hilbert's than to Euclid's, but the subject is still called Euclidean geometry.

Hilbert was working very much in Euclid's tradition. Most of what's in Euclid's *Elements* is mathematical knowledge created by earlier Greeks.[9] Euclid's genius was in gathering all this knowledge together and placing it within a formal system, with clear definitions and postulates. He assembled a huge collection of disconnected results about triangles and circles and conic sections into a logical, coherent whole.

And what Euclid did to Pythagoras, Eudoxus, and others, the professor at Göttingen did to Euclid.

After the definitions and the axioms come the *theorems*. These are the reason for setting up the definitions and axioms in the first place: they are the true statements that follow from the two preceding building blocks. Once Euclid has defined points, lines, and some other things and has set down his assumptions, he can prove, for example, the Pythagorean theorem, a useful fact about right triangles.

Technically, every true statement is a theorem, including $1 + 1 = 2$, but nobody dignifies such trivial statements with the word *theorem*. A theorem is what we call a true statement that is not obvious or that takes some work to demonstrate—and herein lies the beauty of mathematics. The best theorems are surprising. They astonish you, as when you're watching two people play chess, and the game seems routine, and then, suddenly, with a gentle smile, a player slides a piece across the board, and it's checkmate. Nobody saw it coming. The routine theorems are ones that seem as if they must be true but that need to be proven to make sure. The beautiful theorems are statements that you read and think, "Now why should *that* be true?" And then you follow the argument, not seeing where it's going until, suddenly, checkmate. Or, rather, QED.

I've concentrated on geometry as my principal example of an axiomatic system in mathematics because it's the first and, for many, the only place where most students practice the discipline of formal proofs. Geometry is therefore the example most likely to resonate with most readers. Another example, discussed in several later chapters, is the field of mathematics that consumed most of Noether's energies during the later stage of her career: abstract algebra. She not only derived important results in this central edifice of mathematics but also reshaped the entire subject. I mention it here because the essence of abstract algebra is structure, defined by a formal system of definitions and axioms, just as in Euclid's geometry.

Einstein, Hilbert, and Formality

Hilbert was convinced that physics could benefit from a critical examination of assumptions of the kind that he had applied to geometry. He thought that theoretical physicists had a bad habit of patching up their theories by adding on assumptions when they seemed necessary, without checking whether the new assumptions were redundant with the existing ones or even contradicted them. For this unambiguous study of assumptions to even be possible, a physical theory would need to be *formalized*, to be made to look like geometry, with definitions and axioms carefully laid out. This was not the usual style with which physicists built their castles. Einstein, in particular, didn't see the point. He thought that a formal approach to physics was, at least in some cases, a waste of time.[10]

A more modern view might consider Hilbert's delving into gravitational theory a departure from his main concerns, and indeed some physicists of the time (or perhaps only Einstein) might have considered it an intrusion. But to Hilbert, it was just another episode of discovery, of the yearning to satisfy intellectual curiosity about a world that, in his mind, was an organically unified whole.

Einstein's attitude toward highly formal approaches to theoretical physics is expressed in a letter to Felix Klein at the end of 1917: "I am returning to you your second lecture notebook, which I obtained from our colleague Sommerfeld a few days ago, with many thanks. It does seem to me that you are very much overestimating the value of purely formal approaches. The latter are certainly valuable when it is a question of formulating definitively an already established truth, but they almost always fail as a heuristic tool."[11]

Here Einstein is voicing his skepticism of the ability of formal mathematics in helping to discover new physics, while he acknowledges its value in clarifying the structure of a theory *after* the physicists have already groped their way to the essential results. Einstein would retain this skepticism toward formalizations of physics for the rest of his life.

Before developing a keen interest in general relativity, Hilbert had already successfully attacked other problems in physics. For example, he did fundamental work in statistical mechanics—the science of calculating the aggregate behavior of huge numbers of interacting particles.

Relativity already had an established history at Göttingen, long before Einstein got there. The Lorentz transformation, the formula behind the weird and exotic behavior of space and time in special relativity, was derived there first by Woldemar Voigt—a fact not widely known. Göttingen's involvement with relativity continued when Minkowski, who had joined Hilbert there in 1902, gave the 1907 talk in which he introduced the concept of four-dimensional space-time in the context of special relativity.[12] In a subsequent paper, Minkowski pointed out how Newton's theory of gravity was lacking from a relativistic point of view.[13] It's astonishing how quickly Einstein's former math teacher absorbed the new theory and penetrated deep into its formal structure. His four-dimensional rotation model is the way special relativity is often taught and understood today in advanced treatments, although Einstein, at first and for a while, was skeptical of its value, as we saw in Chapter 1.

Then, in 1909, the legendary mathematician Henri Poincaré visited Göttingen to give a series of talks that were an extraordinary precursor to Einstein's visit. But that's better looked at later on.

All this should make it unsurprising that, when Einstein began to lecture and write about his new theory of gravity, Hilbert got wind of it and wanted to know more.

Curved Space

Let's return to Albert Einstein, sitting on a train from Berlin, heading for the college town of Göttingen. The trip takes two or three hours today; back then, in 1915, on the already famously efficient German railway service, it wouldn't have taken much longer.

Say "Einstein" to most people, and a mental image appears: the wise, wrinkled face; the unruly mane of white hair; the international press seeking out his opinion on everything from politics to the existence of God.

But the Einstein on the train was young and handsome. His hair was more or less combed. Although he was well known among physicists by then, nobody on the train would have recognized him. He had nothing of the avuncular old sage about him. In fact, his outlook was tinged with irony, and he was more than a little arrogant. Even the kindly old sage image was largely a creation of imaginative journalists and his earlier biographers. The truth is that he never outgrew his ironic attitude toward what he called the "shabby world" or his impatience with the dull modes of thought characterizing the people infesting it.

He was doing well now, professionally. But his arrogance, as we've seen, had delayed his entry into respectable academic life and was at least part of the reason for his famous exile to the Swiss patent office. It was there, during his few years as a lowly civil servant, that Einstein had most of the important scientific ideas that would ever occur to him. That was his own assessment, made while looking back on his own career. It was at the patent office where he wrote his five amazing physics papers, all published in 1905, the year in which he also found the time to earn his PhD.

Noether obtained her doctorate a few years after Einstein. As we'll see, she also suffered delays and obstacles in her professional career. However, in her case, these roadblocks were not due to arrogance on her part or disregard for the feelings of others. Although her abrupt manner could leave some people taken aback at first, her demeanor was not the source of her difficulties; they were entirely due to the severe sexual discrimination built into German law and custom. The institutional obstacles Noether faced were far more severe than those Einstein had to overcome. And Einstein was eventually to enjoy privileges commensurate with his immense contributions to science, and a reputation in the popular imagination that continued to swell after his death. The injustices suffered by Noether, in contrast, would characterize her entire career in Germany, would follow her to the United States, and would continue to pursue her after her death.

Einstein's amazing 1905 papers spurred his colleagues to coax him out of that patent office and install him as a respected research professor. But Einstein's transformation from someone known only within the tiny prewar community of physicists to the most famous scientist who ever lived—someone whose instantly recognizable face would someday decorate postage stamps and calendars—would be based on the sheaf of papers he clutched, now, on the train. However, the equations on the papers, scribbled in fountain pen in Einstein's rather neat mathematical handwriting, were not working yet.

Compared with the attempt at a theory of gravity that he was traveling to lecture on, his 1905 theory, the work that contained $E = mc^2$, was, in Einstein's words, "child's play."[14] As early as 1912, Einstein had realized that he would need to wield some form of arcane (to him) geometry to properly express the ideas of general relativity.[15] He turned to his friend Marcel Grossmann, a talented mathematician who had been one of Einstein's schoolmates at university, for help.[16] He reportedly wrote to his old friend, "Grossmann, you've got to help me, otherwise I'll go crazy!"

Einstein had sought his help before in studying for his final examinations, for Grossmann had actually attended their math classes and took notes. Now Einstein turned again to his friend and mathematical adviser to search in the library for some textbook or other that might contain the mathematical knowledge needed to describe the way the physicist thought the universe should work. Grossmann found such a book, returning from the stacks with knowledge about recently developed math that seemed tailor-made for their gravity problems. Grossmann's announcement came with a warning that these were deep waters inhospitable to a mere physicist. However, his descriptions of the nonlinear terrors abounding in this complex geometry of calculus on curves and surfaces only convinced Einstein that *here* was the language that he needed.

Although this exotic mathematics was complicated and arcane, it was partly motivated by the concerns of ordinary mortals—after all, the mathematicians who developed it knew nothing of Einstein's revolutionary conception of how gravity operated. However, a calculus that worked on curved surfaces had many other applications, for example, in describing the optimal course that an airplane should take when following the globe from New York to New Delhi. It turned out that the mathematics invented for geometry in a curved space was what Einstein needed because of the way his unfinished theory described the universe, where something like curvature was a property of space itself. His extensive use of this branch of mathematics in his physics turned a circle when one of the authors of the math included chapters on its applications to general relativity in a subsequent textbook about his geometry.

And fortunately for Einstein, Noether was already familiar with at least some of this mathematics and would be able to quickly master whatever pieces she needed to become his final tutor.

A Meeting of the Minds

Hilbert was happy that Einstein had finally accepted his invitation to speak. He had financed the visit the same way he had paid for a series of visits by other scientists and mathematicians to his department: by dipping into the interest earned by the Wolfskehl Prize fund.[17] This was a big award, about a million dollars in today's money, that the mathematician Paul Wolfskehl set up in his will at the turn of the century to be given to the first person to prove the infamous Fermat's last theorem. (There are at least two stories about why Wolfskehl did this, both very dramatic, but in radical contradiction with each other.[18]) Andrew Wiles finally won the award in 1997, with his 129-page proof, but by then, sadly, the principal had shrunk to about thirty thousand dollars because of the turbulent history of the German currency after the close of World War I. Hilbert joked (or maybe not) that he preferred that Fermat's last theorem not be proven, because then the prize would be awarded and his source of funds for visitors cut off. In any case, Einstein's talks during this visit were officially his Wolfskehl Lectures.

Einstein's biographer Abraham Pais sketches a portrait of the scientist and the status of his theory in progress two years before these lectures: "He has no compelling results to show for his efforts. He sees the limitations of what he has done so far. He is supremely confident of his vision. And he stands all alone."[19]

By the time of his 1915 train journey to the great university, Einstein had made some progress with his equations, but the status overall was as Pais describes it. The lectures, however, turned out very well for Einstein, indeed, far better than he had expected.

Einstein gave a crash course in relativity theory for the mathematicians. He spent the first week of July in Göttingen staying in Hilbert's house most of the time and wound up giving six two-hour talks. He began with the special relativity theory of 1905, recounted his long struggle to generalize this theory to include gravity, and ended with his latest, still-incomplete version of his gravitational equations.

An odd twist was that the physicist had to devote part of his lectures to explaining some of the math to the mathematicians.[20] Although Einstein was generally still naive in most areas of math compared with his audience at Göttingen, he had needed to pick up, with the tutelage of his friends, an expertise in some fairly obscure corners of recently developed geometrical methods. They were part of the language he needed for his revolutionary geometrical description of gravity. It was a massive struggle for Einstein to learn this math quickly and thoroughly enough to apply it. As he wrote to physicist Paul Hertz, "You do not have the faintest idea what I had to go through as a mathematical ignoramus before coming into this harbor."[21]

Much as we'd love to know more about the critical conversations between Einstein and Hilbert during the week of their first meeting, we simply do not, explains David E. Rowe, historian of mathematics: "We know almost nothing about what Einstein and Hilbert talked about during the physicist's week in Göttingen."[22]

The arcane mathematics that Einstein had to become proficient in begins with the non-Euclidean geometry that I briefly described earlier in this chapter. In the early eighteenth century (at Göttingen, of course), Carl Friedrich Gauss took the mathematics further into a highly developed theory of curved surfaces. But although the math was already complicated, these were just the familiar two-dimensional surfaces, such as the surface of a globe, that were part of our commonplace, visible reality. Already before 1912, Einstein had realized that general relativity would have to be a theory involving curved four-dimensional space-time. It was this that he had asked Grossmann for help with. The physicist didn't know if such a mathematics existed.

Grossmann discovered the mathematics of Bernhard Riemann and taught it to Einstein. Riemann had developed his ideas about non-Euclidean geometry in multiple dimensions in a PhD thesis at, yes, Göttingen.[23] Gauss was one of his thesis examiners and was very impressed.[24]

The Göttingen mathematicians of 1915 were familiar with Riemann's geometry. In fact, Hermann Weyl (a Göttingen mathematician and a former student of Hilbert's) had written a book on this topic in 1913.[25]

Although Riemannian geometry would ultimately provide the geometrical foundation of Einstein's theory, he needed more. As this was to be a physics theory, geometry alone would not be enough. The language of physics is calculus, the mathematics of change, and of how changes change. Most of the equations of physics, and of all quantitative science, involve rates of change. They're called *differential equations*.

The use of this type of equation to describe physics began with Newton. His famous second law, $F = ma$, equating force with the product of mass and acceleration, is a differential equation, because *acceleration* is the change of velocity over time—and velocity is the change of position over time. Consequently, Newton's differential equation is a statement about the change of a change. This form of the equation is the way it's usually taught to physics students, although Newton's original formulation equated the force with the time change of *momentum*, the product of mass and velocity. This version is in fact a more general and powerful form of the second law because it applies directly to situations where the mass changes, such as when a rocket burns off fuel.

Newton's innovation in describing physical reality with differential equations was huge. Einstein described the development this way: "to enunciate the Laws of Motion in the form of total differential equations—perhaps the greatest intellectual stride that it has ever been granted to any man to make."[26]

Newton's space was what we call *flat* today. The change in velocity in his differential equation was any alteration of the velocity vector: a change in speed, of course, but also any deviation from the geodesic in flat space—any variation from the straight line that was the shortest distance between two points. Newton's differential equations, brilliant as they were, only spoke about changes in flat space.

Einstein's space was not flat, and his geodesics were not Euclidean lines. He needed some language in which he could write down *his* differential equations in curved space.

Einstein needed to do calculus in curved, four-dimensional space-time. Here, too, Grossmann found the appropriate mathematical literature, which he and Einstein explored together. This was the math that lets you do physics within Riemannian geometry. One of its central objects is the *tensor*; this branch of mathematics is often called *tensor calculus*, though it is also sometimes called *Ricci calculus* after one of its developers. This form of calculus was developed not by Göttingen mathematicians, oddly enough, but by two Italians, Tullio Levi-Civita and his teacher Gregorio Ricci-Curbastro. In 1915, tensor calculus was a recent development in math. Levi-Civita and Ricci-Curbastro were still alive, and, in fact, Einstein corresponded with Levi-Civita about relativity. Einstein became a real expert in this new math; after a few years, he knew far more about it than did almost all mathematicians, aside from its inventors. It was the new tensor calculus that Einstein found himself explaining to the mathematicians at Göttingen.

———

I'll explain a few simple mathematical concepts to provide some familiarity with the terminology and to help you to appreciate what the scientists and mathematicians were struggling to accomplish. Understanding a few technical terms in the way mathematicians use them will help avoid becoming lost in generalities.

One important object throughout all of mathematics is called the *function*. This is a map, or a rule, that associates numbers with other numbers. If you deposit some money in a bank account that earns some constant interest rate, and the interest earned is deposited back into the account, you may know that the money will increase over time exponentially, because of the effect of compounding interest. If you make a table of how much money you will have each month, the table is the representation of a function: it associates one set of numbers (the months) with another set (your account balance). The table is a sampling of a continuous function that maps time to money.

Calculus is the study of how functions change. (Rather, one of its branches, *differential calculus*, is that study). It can tell you how much faster your money will grow at a higher interest rate.

A *vector* can be visualized as an arrow in space, in one, two, three, or (and here visualization becomes difficult) more dimensions—even infinite dimensions. Crucially, vectors have a reality independent of the particular numbers used to describe their coordinates. If you've used a magnetic compass to navigate, you're familiar with the idea. The needle will always point north; it's a vector pointing in a particular direction. As you slowly rotate the compass, the numbers under the arrow move, but the arrow does not.

The concept of a *tensor* combines the concepts of function and vector. Just as a function is a map from numbers to numbers, a tensor is a map from vectors to vectors. It's a function of vectors that yields other vectors. Technically, to qualify as a tensor, the map must satisfy certain conditions, most of which I won't mention but one of which I will. Just as vectors have a reality independent of the coordinate system, so do tensors. They have to be independent: since we can represent our vectors with any number of different sets of coordinates, it wouldn't make sense for our vector functions, our tensors, to change as we changed the coordinate system. If they did, the idea of a function of vectors would make no sense. Finally, just as calculus is the math of how functions change, tensor calculus is the math of how tensors change.

Einstein was using tensors to describe the four-dimensional space-time that he had invented, and he needed tensor calculus to describe how the physics of gravity worked according to his theory. Using this flavor of math to describe physical reality was a bold and risky move. Two years earlier, Einstein had shown his work to Max Planck, his friend and the physicist who himself had taken the first theoretical step that led to quantum mechanics. Planck had tried to warn him away from using exotic geometry. "I must advise you against it," he wrote to Einstein, "for in the first place you will not succeed, and even if you succeed no one will believe you."[27]

Except that they *did* believe him.

There were a few holdouts in the Göttingen community, those who either did not accept the theory or didn't like the way that Einstein presented it. Klein, for his part, remarked that Einstein was not fundamentally a mathematician but instead seemed to rely on obscure, partly philosophical influences that were at least somewhat mysterious to the venerated mathematics professor.[28]

But overall, Einstein killed it in Göttingen. He convinced most of the important people there, including Hilbert, Weyl, and, despite his bemusement, even Klein, that his theory made sense. They accepted it as a theory, as *the* theory, of gravity.

Return to Berlin

Hilbert was certainly convinced. He had invited Einstein because of his own intense interest in physics—an interest not shared or even understood by some of his colleagues.[29]

Felix Klein was also an immediate convert to Einstein's theory. His endorsement was fortunate because his voice carried tremendous weight among mathematicians and even physicists. Like Hilbert, Klein was immediately fascinated with what would soon be general relativity, but he didn't fully understand it. There were places where the math was unclear to him and not easy to interpret. However, after Einstein returned to Berlin, Klein became one of a small group who carried on a correspondence with him, exploring avenues of development and helping him along on the path to completion.

How did Klein make the jump from unsure observer to participant at the cutting edge? He explained it himself sometime later: it was "Fräulein Noether" who got him up to speed. And during this time, Noether also corresponded with Einstein, who expressed his wonderment at what happened to his equations in her hands, how he never imagined that things could be expressed with such elegance and generality. He even teased Hilbert in a letter, suggesting that the men of the Göttingen math department could have learned a lot from her had they brought her there sooner.[30]

Unfortunately, we have little record of Einstein's twelve-hour seminar on relativity. Some notes on the first lecture have been discovered, but nothing on the following five.[31] And these notes contain no attendance record. Therefore, history does not record whether Noether attended the lectures or whether Einstein met her during his first trip to Göttingen. As mentioned earlier, when her mother had died unexpectedly soon after she arrived at Göttingen, she was obliged to make frequent journeys back and forth and to spend much of her time for some months at Erlangen.

It seems unlikely that Einstein encountered Noether during this visit. In his letters to friends recounting his experiences there, he does not mention her. If he had seen her, it's a sure thing she would have made a strong impression and he would have had some tales to tell.

It would have been impossible not to notice Noether's presence at Göttingen. Even today, a woman among the faculty of a university mathematics department is a somewhat unusual sight.[32] In 1915 Germany, there were no female professors at all. Noether was the only living woman in all of Europe with a mathematics PhD, and she possessed one of only a small handful of such degrees that had ever been earned by a woman.

Surely Einstein would have noticed a woman with such unconventional dress and who paid such scant attention to the arrangement of her hair and clothing. Undoubtedly he would have remarked to his friends about the singular creature he encountered—someone who laughed and spoke loudly, caught on to everything instantly, and did not hesitate to interrupt when she had a good idea. This is the image we've gleaned of Emmy Noether at that time. But as we will see, she and Einstein got to know each other later.

We do know that Hilbert immediately asked Noether for assistance with problems related to general relativity. Her description of this work supplies more evidence that she had been unable to attend Einstein's talks, because, at this stage, she seemed to be unaware of the context for the calculations. Soon after Einstein's lectures, she wrote home about the subject of these calculations, saying that "none of us understands what they are for."[33]

She certainly became aware of what they were for in short order, however, as she began working with Einstein directly. The physicist wrote to Hilbert, praising her assistance: "You know that Frl. Noether is continually advising me in my projects and that it is really through her that I have become competent in the subject."[34]

In any case, the unpaid assistant from Erlangen thoroughly and quickly absorbed all the details of Einstein's theory. We know this because of the continuing correspondence that followed the lectures, Klein's testimony, and what she would soon do. That will be the subject of the next chapter.

Einstein's equations, as I've noted, were not quite finished. There were two problems with his theory as it stood in July 1915. One Einstein knew about and had given up on. The other he didn't seem to realize that he had; or, perhaps, his obsession with the first problem left no time or energy for other details.

The first problem, the one that he had spent most of his time on during the preceding eight years or so, was the issue of *covariance*. This is the property of a set of equations that they be independent of the coordinate system used to measure time and space. In other words, they had to be in a sense objective, not changing form depending on your point of view. This property is related to the invariance of vectors, as in the metaphor of the compass needle, and of tensors, as alluded to earlier. But here it's an issue of the status—not of a particular mathematical object but of an entire system of equations—of the theory as a whole. At least at the outset of his journey, Einstein's philosophical stance was that the theory of gravity should describe a fully objective reality with equations that remained invariant from the point of view of *any* reference frame, including ones that were rotating and accelerating. He called this property *general covariance*, which we'll call *covariance* for short.

Einstein had managed to formulate special relativity so that it was perfectly covariant—its equations were totally independent of the choice of inertial reference frame—and for many years he had thought that he could do the same for his theory of gravity. Once he could manage that, then the gravity theory would truly become a general theory of relativity. But by the time he arrived at Göttingen in July, he had given up. He presented his theory as a kind of compromise, saying that it was as close at it could get to full covariance. It was still workable as a theory of gravity, and you could still make predictions from it. But it would break down in some situations, and some of the predictions would be wrong.

The most significant thing that happened, scientifically speaking, during Einstein's week at Göttingen was that the mathematicians there, Hilbert especially, thought that he was wrong to give up and convinced him of their belief. In the few days they had, they couldn't see exactly how to achieve covariance, but they were sure that it was just a matter of adjusting some of the details and that Einstein was on the right track. This support was another source of elation for Einstein: not only did these folks agree that he had nailed down what a true and beautiful theory of gravity should look like, but they also convinced him that his dream of a true general theory of relativity was within his grasp. Hilbert, especially, with his almost-supernatural ability to intuit the essence of things immediately and his familiarity with every branch of mathematics, watched as Einstein wrote his equations on the board. The mathematician saw things in them that Einstein could not. Einstein would soon learn just how far Hilbert's gaze had penetrated.

Einstein's ability to persuade at Göttingen was no small thing. Up to then, he had found few colleagues who could understand what he was doing; now he was nearly ecstatic. Immediately after returning home, he wrote joyful letters to his friends:

[To Heinrich Zangger, July 7, 1915]

I held six two-hour lectures there on gravitation theory, which is now clarified very much already, and had the pleasurable experience of convincing the mathematicians there thoroughly. Berlin is no match for Göttingen, as far as the liveliness of academic interest is concerned, in this field at least.[35]

[To Arnold Sommerfeld, July 15, 1915]

In Göttingen I had the great pleasure of seeing that everything was understood down to the details. I am quite enchanted with Hilbert. That's an important man for you! I am very curious to know your opinion.[36]

[To Heinrich Zangger, between July 24 and August 7, 1915]

Never have I been able to rest so well since adulthood. Surely I have already written you that I held 6 lectures at Göttingen in which I was able to convince Hilbert of the general theory of relativity. I am quite enchanted with the latter, a man of astonishing energy and independence in all things.[37]

Einstein might not even have realized what an accomplishment it was to convince Hilbert that he himself had gotten it right. After all, the outspoken mathematician had recently cracked wise that he thought "physics is much too hard for physicists."[38]

Einstein's appreciation of Hilbert was not due solely to the famous mathematician's scientific acumen. The two men spent a good amount of time during the week just hanging out and talking about politics and the war. Einstein felt he had found a kindred spirit, one who also sought to remain above the pointless squabbles among nations and to keep the focus on truth and beauty. He was happy to have found a new friend: "But in these times," he wrote to Zangger, "you appreciate doubly the few people who rise well above the situation [politics, war] and do not let themselves be carried along in the turbid current of the times. One such person is Hilbert, the Göttingen mathematician. I was in Göttingen for a week, where I met him and became quite fond of him."[39]

Although Einstein and Hilbert shared a general political outlook, as well as the courage and independence required to do something with it, then and for at least the next few years Einstein maintained a certain political naivete in which Hilbert, with his comparative shrewdness, did not partake. This difference had a consequence in April 1918, when Einstein, remembering Hilbert's political independence and the reputation that had borne it out, wrote to him about his idea for a book.[40] The volume would contain essays by prominent people of the arts and sciences against nationalism and the war. He asked Hilbert for his opinion of the project and whether he would be interested in contributing an essay. In a follow-up note, Einstein listed the names of others he had sent the proposal to and remarked that "no one whose name has been tainted by chauvinistic declarations of any sort is being invited." He added, "Could you perhaps name a few suitable persons to whom I could also turn concerning this matter? Among mathematicians and physicists in Germany you are unfortunately the only one to whom I can write about this affair."

Hilbert replied to Einstein a few days later, saying that he thought the idea was not a good one, for tactical reasons, and declined to participate.[41] He also noted that the few other possible essayists whom Einstein had adduced were not good choices, for a variety of reasons.

Einstein eventually gave up the idea, writing back to Hilbert near the end of May 1918 about his disillusionment with the others who he had hoped might join in the project.[42] He revealed in this letter that he had envisioned his project as an "antithesis" to the notorious German cultural declaration that Hilbert had previously refused to sign.

Perhaps this episode was a well-timed lesson in political tactics for the young physicist. On the near horizon was something that would create, for him, circumstances that he could never have imagined in 1918.

Soon everyone would want to know his opinion about everything and would take it seriously.

But now, for a few days near the end of 1915, Einstein was back in Berlin and enjoying a relaxed satisfaction that had eluded him in his professional life for a very long time.

The Race

Then Hilbert dropped a bomb that plunged Einstein into a desperate frenzy of work that nearly destroyed his health.

Einstein had taken Hilbert's assurances to heart. He now believed that he could massage his equations into correct covariant form and create what he had dreamed of for so long: a true general theory of relativity.

And it would belong to him. Although many people had helped him, and would help him, arrive at a final set of equations, he had no intention of sharing any of the credit, as he made perfectly clear in his letters. For example, he told physicist Arnold Sommerfeld, "Grossmann will never lay claim to being co-discoverer. He only helped in guiding me through the mathematical literature but contributed nothing of substance to the results."[43] Grossmann's name, however, does appear as coauthor on several of Einstein's gravitation papers before 1915, as they were working together to try to construct a covariant theory.

As Rowe puts it, "After he abandoned the Entwurf theory, Einstein seems to have mentioned Grossmann's name with but one purpose in mind: to secure his priority as the discoverer of the correct gravitational field equations."[44] The so-called Entwurf (preliminary or draft) theory refers to an earlier version of general relativity before general covariance was achieved, expounded partly in Einstein and Grossmann's joint papers.

This view of a jealous and possessive Einstein may not accord with the popular image of the altruistic sage, but when it came to general relativity, his grasp was tight. Einstein seemed to understand the historical importance this theory would hold, and he needed it to be his alone.

Of course, Einstein not only omitted Grossmann's name from any papers on general relativity from 1915 onward but also never felt moved to include Noether's, either on an author list, in a footnote, or in the acknowledgments, despite expressing gratitude for her tutelage in his correspondence. Although this omission is unfortunate, we should nevertheless remember that the cultural standards and practices around scientific publication, and the apportionment of credit, are today considerably different from how they stood in the early to middle twentieth century.

While working on general relativity, he was exchanging letters with Hilbert. We can try to imagine Einstein's state of mind when he received a draft of a paper from Hilbert with the mathematician's own version of the theory of gravity. In some ways, Hilbert's was nicer than Einstein's version.

The central object in Einstein's formulation of general relativity was the tensor that described the shape of space-time. The theory's equations described how this tensor took its form from the distribution of mass and energy in the universe. In the four-dimensional space-time of general relativity, the tensor had ten independent components; consequently, Einstein's formulation consisted of ten equations.

Hilbert, the consummate mathematician, had found a way to replace Einstein's ten complex equations with a single, more beautiful equation, from which all the details could be extracted.

Einstein's mood and his attitude toward Hilbert suffered a rapid deterioration. He writes to the same friend, Heinrich Zangger, with whom he had shared, in July, his joy at getting to know the mathematician: "The theory is beautiful beyond comparison. However, only one colleague has really understood it."[45]

The colleague he refers to is, of course, David Hilbert. But, he continues: "and he is seeking to *partake* in it . . . in a clever way. In my personal experience I have hardly come to know the wretchedness of mankind better than as a result of this theory and everything connected to it. But it does not bother me." The emphasized word *partake* is an attempt to translate an expression that has a certain ironic meaning in the original. Here Einstein refers to Hilbert's effort to subsume general relativity into a wider theory of his own, one that would include electromagnetism, in an early attempt at a modern unified theory. In any case, Hilbert's try at a unified theory didn't work and is now merely of historical interest.

Einstein's complaint was not an isolated outburst. A few days after his letter to Zangger, he also wrote to his close friend Michele Besso: "My colleagues are acting hideously in this affair."[46] (It's Hilbert again.)

The mathematically adept engineer Besso might, along with Hilbert, Noether, Klein, and Grossmann, be fairly considered one of the authors of general relativity.[47] He had also given Einstein critical help in understanding and applying the advanced geometry that he needed to describe the structure of space-time. Besso is one of several people whose role in helping to give birth to the theory remains uncredited in almost all popular accounts. (Recently, a manuscript written by Besso and Einstein between 1913 and 1914, with about half the pages in Einstein's handwriting and half in Besso's, sold at auction for thirteen million dollars. This is a record amount for the sale of an autographed scientific document.[48])

Feeling the theory slipping from his grasp, Einstein set to work as he had never done before. His obsession with work was extreme, even for him. He neglected eating, sleeping, and everything else aside from his calculations. Things culminated in November 2015, when he presented four papers to the Prussian Academy on four consecutive Thursdays. The audience could practically see the baby being delivered in real time. In the third paper, Einstein calculated the correct orbit for Mercury, something that had been a mystery since Newton's time. This calculation was already an astonishing, historic achievement. The fourth and final paper dotted the final *i* and crossed the final *t*. The general theory of relativity was complete.

Einstein was exhausted, his body nearly broken.

Almost simultaneously, a paper by Hilbert appeared with a complete, fully covariant theory of gravity. The paper was equivalent to Einstein's but used the mathematician's elegant methods.

To this very day, historians are debating who should be awarded priority and how credit should be divided. Archival research and intrigue continues, with regular arguments about the meanings of scribbled notes and the authenticity of printer's proofs. These avid investigators might consider taking a clue from the great man of Göttingen.

As soon as Hilbert saw that the question of priority was about to become an issue, he put a stop to it. He let it be known that this was Einstein's theory and that Einstein should get the credit. For even if Einstein's mathematical formulation was not the most sophisticated, the original idea—the physical idea that gravity was a result of a curved space-time—belonged to him. As Hilbert put it, "Every boy in the streets of Göttingen understands more about four-dimensional geometry than Einstein. Yet, in spite of that, Einstein did the work and not the mathematicians."[49]

His amusing exaggeration aside, it is impossible to know what Hilbert was actually thinking. Did he really believe that Einstein deserved all the credit for this revolutionary theory of gravity? Whether Hilbert was able to foresee the judgment of history or not, history has judged him well. He gets credit for being a scholar and a gentleman, and among those who have looked deeply into the events of the second half of 1915, he also gets credit for playing a perhaps indispensable part in birthing general relativity.

The idea of a race to reach the fully covariant equations of general relativity was most likely a notion that existed, to the extent that it existed at all, mainly inside Einstein's mind. The relevant correspondence, sparse as it is, supports this view, despite the excited recounting of this so-called race by a handful of subsequent chroniclers. The actual evidence suggests free and generous help and sharing of ideas by Hilbert, Klein, and Noether. It is doubtful that without their help Einstein would have been able to straighten out his math and reach his result at nearly the same time as Hilbert, or perhaps ever. For this reason, I believe that these three and at least Grossmann and Besso should be regarded by history as coauthors of the great theory. But ultimately, squabbling over the fair shares of credit is beside the point. Were we not encouraged by a jealous Einstein himself to pay attention to these matters, later writers would probably have had less to say about them.

If I have an interest in a fair accounting of contributions, it's only in elevating the share accorded to Emmy Noether. A few scientific historians do point out the role played by Klein in working with and alongside Einstein and Hilbert during the genesis of the field equations, and the value of his work in the years after 1915 on questions related to general relativity is widely acknowledged. However, Klein's own statements make it utterly clear that he would have been unable to perform any of this work without the tutelage and active assistance of Noether. Shortly after the "race" had ended in its photo finish, he reminded Hilbert of her role, in the context of his ongoing work in gravitation: "You know that Miss Noether continues to advise me in my works and that it is only because of her that I was able to delve into the present material."[50]

He acknowledged her contributions to his gravitational and related work in several other places as well, but if we overlook this commentary and other clues to the development of ideas, we could easily form an inaccurate impression of the credit that we should assign to various people. Modern scientists are particularly susceptible to falling prey to such a distorted picture because we pay disproportionate attention to the lists of authors on papers and forget, or may be unaware, that the conventions of formal authorship are highly dependent on time and place. If we transported Klein to an American university and to the twenty-first century, papers that he had published as the sole author in 1919 would no doubt now include Noether as a coauthor. Here, as in so many places, we should strive to avoid anachronistic judgments. Klein had no intention of suppressing Noether's role; indeed, he was instrumental in promoting her at every opportunity. It is rather that the conventions of authorship have changed. Papers from this era are often accounts of work performed by entire laboratories of scientists but bear the name of the single prominent member of a scientific society who happened to present the report.

And there is Noether herself, who cared nothing for self-promotion and who casually gave away results to students and younger colleagues, and who, after discovering her eponymous theorem and describing it in her elegant and dense paper, displayed no further interest in any part of physics. But she was happy to lend Klein, for whom she felt gratitude, friendship, and, like everyone else, a certain reverence, any help that he needed.

Before leaving, for the moment, the subject of the credit allotted to Noether in the early development of gravitational theory, we must confront the apparent likelihood that certain authors suppressed or omitted her contributions where it would have been natural to do the opposite. One example closely related to the present subject is the case, which will be painful for physicists to contemplate, of Wolfgang Pauli. The Austrian physicist is universally revered as a theoretician; among other accomplishments, he formulated the Pauli exclusion principle in quantum mechanics. This principle provides the ultimate explanation for the organization of the periodic table and thus the basic realities of chemistry. In 1918, Pauli applied a different sort of exclusion principle when writing an encyclopedia article on relativity. In comments on a draft, Klein sent him references to directly relevant papers by himself, Poincaré, Hilbert, and Noether. In Pauli's final version, he included mentions of all these works except Noether's.[51]

Returning to Einstein, there is also some evidence of direct communication between him and Noether during this time, but most of her insight was conveyed to him by Klein as an intermediary. Klein and Noether were, after all, in the same place at the same time, enthusiastically immersed in the Göttingen tradition of mathematical walk and talk. (Klein shared Hilbert's and Noether's love for mathematics on the move. As a young professor at Erlangen, he had written to Emmy's father, Max, "If only I had someone here with whom I could talk reasonably about things. I understand matters very well when I can question someone about them, but hardly at all when I am staring at the printed material in front of me."[52]) And it seems that, at this stage and to some degree forever after, Emmy Noether looked at these problems of Einstein's as mathematical problems. She never developed the interest in physics as physics that Klein certainly had. But as we will see in the next chapter, she certainly understood the physical content of the work and the physics implications of her great result that came out of it.

Klein gave his own opinion about the issue of priority a few years later: "The question of precedence is not an issue, because both authors [Einstein and Hilbert] follow completely different trains of thought (so much so that their results did not seem compatible at first)."[53] It was nearly a year after these first publications of the gravitational equations that Einstein himself was to show that the two approaches were equivalent.

There is a final irony that Hilbert didn't know about and that Einstein himself may have forgotten. Einstein had collected the pieces of his puzzle, all the components of the final, fully covariant equations, two years before, in one of his notebooks.[54] He had even worked out the precession of Mercury's orbit.[55] However, he had abandoned this work and begun anew, somehow not realizing that he was in sight of his holy grail.

Einstein's 1915 visit was reminiscent of a visit to Göttingen in 1909 by Poincaré, three years before his death.[56] The French mathematician came to talk about relativity and about the wonderful and surprising behavior of space and time.

Poincaré was a brilliant, legendary mathematician, mathematical physicist, and philosopher of mathematics. His connection with

Einstein's relativity theories, both special and general, is rather curious. In some ways, he anticipated aspects of both theories but never managed to make the bold leaps in physical thought that Einstein made. Consequently, despite having in his hands many of the theories' equations, he did not transform this possession into the revolution that we owe Einstein.

Poincaré also presented a series of six talks to the Göttingen group. His talks on the "new mechanics" shared much with the formal content of Einstein's special relativity (to which he never referred!), but Poincaré failed to grasp both the theory's physical significance and crucial elements of its logical structure. These aspects Einstein, with his comparatively limited mathematical toolbox, had penetrated into, all the way to their astonishing cores, through his relentless process of thought experiment combined with his fearless acceptance of its counterintuitive results.

The story of Poincaré's involvement with relativity gives us another example of Einstein's "doing the work" while surrounded by mathematicians who were better acquainted with the language of that work than he.

The general theory of relativity is an example of a *field theory*. In physics and mathematics, a *field* is a function that maps every point in space (or space-time) to a number or a vector. If you've seen a weather map, you've seen examples of visualizations of fields. Red for hot, blue for cold, the colors overlaid on part of the surface of the earth—this is a temperature field. The array of little arrows showing the intensity and direction of the wind is a representation of a vector field. If you can imagine a map of the temperature everywhere in your room, this would be a *three-dimensional* temperature field.

In the previous chapter, I mentioned that one of the conceptual problems with the existing theory of gravity was the instantaneous action at a distance that it entailed. In Einstein's world, nothing, including the force of gravity, could travel faster than the speed of light, and this was one of many things that general relativity took care of.

Field theories treat the constituents of physical reality as being spread over space. Newton's theories of motion and gravity were not field theories but instead were theories of separate objects interacting over distance through the instantaneous transmission of forces. To a certain kind of physicist mind, this model was unsatisfactory. Something over there could not influence something over here instantaneously; it seemed too mysterious. The influence needed to travel over the intervening space, and its propagation through that space, should be part of the theory that described the interaction. If someone a mile away speaks to us, we can't hear the message until the sound travels from there to here; a reasonable theory should be able to describe how the atmosphere allows the sound to propagate. Like all field theories, it should be *local*, meaning that it should account for how the sound travels from each location in space to the next location just a bit closer to us, eventually covering the entire distance. Field theories of gravity or light should have the same character, even if the signals they describe travel through a vacuum.

An example of a successful field theory, one whose structure guided and inspired Einstein, was Maxwell's unified theory of electricity and magnetism. We'll call this theory *electrodynamics* for short. In fact, one of the motivations behind special relativity was the observation that Newton's mechanics was incompatible with electrodynamics. In electrodynamics, things transform according to the Lorentz transformation, the transformation law for which Minkowski found an elegant description as a rotation in four dimensions. Newton's mechanics, however, obeyed Galilean relativity. The two are not compatible, although they become indistinguishable at velocities much smaller than the speed of light. Electrodynamics, the nineteenth-century theory of electricity and magnetism, manifested, in other words, all the weird phenomena that

Einstein described in his theory of relativity: clocks ticking at different rates and rulers contracting. By some accounts, electrodynamics ushered in the beginnings of modern physics. In 1905, Einstein was faced with two theories, both apparently successful but obeying two different transformation laws that could not be reconciled. In Newton's universe, all clocks measured time at the same pace and rulers could be relied on: an inch was always an inch.

Einstein's genius allowed him to find a way to reconcile mechanics with electromagnetism. He explained how the Lorentz transformation applied to masses in motion as well as electric and magnetic fields, including the propagation of light. This was the theory of special relativity, which introduced a new conception of space, time, mass, and energy. The theory explained why the speed of light was the same for all observers. It did so by including the light speed in the formula that transformed the velocity of material objects as the observer changed reference frame.

The general theory, like electrodynamics, was a field theory, and, like electrodynamics, its influences propagate through space at the speed of light rather than acting instantaneously. In field theories, the mathematical construct of the field takes on a life of its own. No longer do forces act directly between objects, whether charges, like electrons or protons, in electrodynamics, or masses, in general relativity. Instead, the objects act on the space-filling field, which in turn acts on other objects.

Einstein's difficulties in making sense of a field theory of gravity had a fascinating precursor, if perhaps a faint one, in a digression by Scottish physicist James Clerk Maxwell. Maxwell was the creator of the first great unified field theory in physics. He pulled together the experimental and nascent theoretical threads running through the disparate phenomena of electricity and magnetism—the investigations of André-Marie Ampère, Michael Faraday, and others—into the elegant synthesis of Maxwell's equations. These four (or two or one, depending on how fancy your math is) equations assembled the assorted phenomena of motors, generators, and all the other fascinating behavior teased out of the ozone-filled steampunk laboratories of the nineteenth century into a timeless, crystalline mathematical structure. This was the first real field theory in physics: a model of a world not with bodies acting mysteriously and instantaneously at a distance, as did Newton's planets and fruit, but with interactions mediated by continuous fields that permeated the space between them.[57]

Einstein attached enormous significance to Maxwell's employment of the field concept: "This change in the conception of Reality is the most profound and the most fruitful that physics has experienced since the time of Newton."[58]

But when Maxwell turned, for a moment, in the middle of his seminal monograph on his electromagnetic theory, to see what would happen if he applied his ideas to gravity, he found his way stymied by paradoxes, or at least by unacceptable conclusions. He discovered that the peculiar nature of gravitational interactions, that like "charges" attract rather than repel as in electromagnetism, led to the conclusion that the gravitational field must possess an "enormous intrinsic energy." Maxwell then sadly concludes, "As I am unable to understand in what way a medium can possess such properties, I cannot go any further in this direction in searching for the cause of gravitation."

Maxwell's perplexity foreshadows Einstein's troubles in constructing a consistent field theory of gravitation. The Scottish physicist's abandonment of the program, when faced with questions of *energy* in the field that didn't seem to have reasonable answers, will have an echo in the next chapter, when we reach the part of our story where Emmy Noether tackles the looming issue of energy and gravity.

The hard, years-long work of finding generally covariant field equations, of achieving a general theory of relativity, was finally done. But in another sense, Einstein's work was just beginning. He had other scientific interests, which he continued to pursue, but our story mainly concerns the fate of his new theory of gravity.

It took eight years of unceasing labor and the help of some talented friends and a few of the greatest mathematicians in the world before Einstein was able to write down the final version of the gravitational field equations. As he wrote to his colleague Rudolph Förster near the beginning of 1918: "You must not take your industrial slavery too much to heart. I also was one of those during my best years, as an employee of the Swiss Patent Office. My genuinely original ideas all stem from that time, for all too soon does one become an avuncular old boy, and the little fount of imagination dries up."[59]

Einstein was keenly aware that general relativity was quite unlike anything that had come before. In addition to its inherent radical nature, it depended intimately on mathematical techniques that were not taught routinely to physicists. The only physicists likely to have had contact with this brand of tensor calculus were those who were also mathematicians, and not many of them, and any mere physicists who were to learn the theory would need an unusual mixture of strong motivation and highly developed mathematical ability. That's why Einstein was so delighted when he encountered Hilbert and a few of the other Göttingen mathematicians who caught on quickly and believed in his ideas. He obviously did not expect such receptivity, and he felt supremely validated.

But now Einstein desired to promote his theory in the wider universe of physicists. Indeed, in his letters it is easy to detect not only an eagerness that others learn of and try to understand his work but also an apprehension that they might reject it, either out of ignorance, incompetence, or just the inability to absorb and appreciate what was, after all, in the words of Walter Isaacson, a "whole new way of regarding reality."[60]

Einstein's state of mind in the aftermath of his (and Hilbert's) publication of general relativity can be gleaned from a note he wrote to Hermann Weyl about a year later: "I am extremely pleased that you have welcomed the general theory of relativity with such warmth and enthusiasm.

Although the theory still has many opponents at the moment, I console myself with the following situation: The otherwise established average brain power of the advocates immensely surpasses that of the opponents! This is objective evidence, of a sort, for the naturalness and rationality of the theory."[61] In other words, the smart people were on his side.

Einstein kept an eye on how his theory was faring beyond Germany, especially as the war was interfering severely with communication and the dissemination of ideas.[62] Near the beginning of 1917, he wrote to the physicist Willem de Sitter: "It is a fine thing that you are throwing this bridge over the abyss of delusion," referring to de Sitter's efforts to spread knowledge of general relativity among British scientists.[63] (De Sitter did more than just advocate for general relativity. He found one of the first cosmological solutions to Einstein's equations, solutions that describe a symmetrical expanding or contracting universe.)

Felix Klein was also keeping an eye on things, as usual, but his point of view was somewhat different. He was concerned about nurturing the reputation of his beloved mathematics department and the Göttingen mathematicians. In particular, although he greatly admired Einstein and his accomplishment, he began to be a bit put out that people were ignoring Hilbert's role in particular. Klein wasn't even overly fond of Hilbert's approach to physics, especially how it had manifested itself in his formulation of general relativity. Hilbert's elegant equation, mentioned earlier, was a *variational principle*, something that I'll explain in the next chapter. Klein was not on board with Hilbert's "fanatical belief in variational principles."[64] He even had a larger objection to Hilbert's stance that "the essence of nature could be explained by mere mathematical reflection." But as general relativity became more famous, Klein still wanted the Göttingen crowd to get a larger share of the credit. In a letter to Wolfgang Pauli, Klein pouted that the "physicists mostly pass over Hilbert's contributions [to gravitation] in stony silence."

(Einstein himself had tried to derive his gravitational equations from a variational principle at least as early as 1914, and again during the dramatic month of November 1915.[65] However, because he was deriving the wrong equations—he didn't have the right ones yet—this exercise led nowhere.)

Einstein harbored no delusions that he had found a timeless truth. He just felt that his theory was extremely beautiful, which it was, and that it was the best, the only currently viable, description of gravity and the large-scale structure of the cosmos. Of course, he realized that just as his theory of gravitation had supplanted Newton's, one day the theory of general relativity, at least the version with Einstein's name on it, would inevitably be supplanted by a theory as yet unknowable. "I believe that this process of securing the theory has no limit," he wrote.[66] By this statement, Einstein means that there will be no final theory but rather an unending succession of models achieving ever greater mastery over nature.

Although he had grounded his physics ideas in philosophical precepts, Einstein was uncomfortable with speculative philosophy and perhaps metaphysics in general: "From reading philosophical books I had to learn that I stand there like a blind man before a painting. I grasp only the inductive method at work . . . but the works of speculative philosophy are inaccessible to me."[67]

While, during these months, Einstein was excited about his scientific successes and eager to see his new theory gain wider recognition, he was also distraught by the unceasing drumbeat of horrible news about the European war: "It is a pity that we do not live on Mars and just observe man's nasty antics by telescope. Our Jehovah no longer needs to send down showers of ash and brimstone; he has modernized and has set this mechanism to run on automatic."[68]

It was not merely the fighting and destruction but also the effect of mindless nationalism on the minds of the people around him that disturbed him: "Politically, things look strange. When I speak with people, I sense the pathology of the general state of mind. The times recall the witch trials and other religious misjudgments. It is precisely the responsible and privately most selfless of people who are frequently the most rabid supporters of dogmatism. Social sentiment has gone badly astray. I could not imagine such people existed, if I did not see them before me. Now I can only hope for salvation from an external force."[69]

These sentiments foreshadow even worse times to come—times that will directly affect the lives of Albert Einstein, Emmy Noether, David Hilbert, and all their friends and colleagues.

To get the physics community really interested in general relativity, Einstein had to show that it was physics and not just a mathematical dream. He also had to prove this to himself.

Physics is an empirical science. Its theories are no good unless they not only agree with what we've already observed but also make *new* predictions about things not yet seen.

There is a popular image of Einstein as a solitary, theoretical toiler, perhaps even disdainful of experimental evidence and contemptuous of practical matters. This impression is, to be fair, reinforced by a handful of comments uttered by Einstein himself, some genuine, some perhaps apocryphal. He probably *did* say that he would have been "sorry for God" had certain astronomical observations (described in Chapter 7) failed to confirm the predictions of his theory of general relativity because "the theory is correct."

Whether you believe that Einstein thought experimental agreement important or irrelevant, you can find plenty of Einsteinian pronouncements to support either view. Looking at everything he said on the matter, he was most likely not equivocating. He probably entertained both points of view at the same time.

Einstein was a physicist, not a mathematician. He conducted experiments himself and enjoyed them. He invented things and held patents. He kept himself aware of experimental developments in the literature and was guided by them.

As he developed versions of his theory of gravity over the years, he kept returning to the calculation of the orbit of Mercury as a check on the status of his theory. When the predictions failed to agree with observation, he accepted these findings as evidence that something was wrong with the theory.

In a 1919 letter to the *London Times*, he discussed possible experimental tests of general relativity: "The chief attraction of the theory lies in its logical completeness. If a single one of the conclusions drawn from it proves wrong, it must be given up; to modify it without destroying the whole structure seems to be impossible."[70]

Newton's gravity, which seemed to predict everything else about the universe correctly, also predicted that Mercury's orbit would be an eternal, closed ellipse, just as with all the other planets. However, deviations from this prediction had been bedeviling astronomers for a long time, as described in the previous chapter. The deviation was usually described as a precession, or gradual rotation, of the *perihelion* of the orbit: the shortest distance from the planet to the sun.

Most astronomers assumed that the source of Mercury's precessing perihelion was the perturbing effect of an as yet undiscovered planet. In fact, some of the other planets had observably imperfect orbits, and those imperfections had been traced to the interference of the other inhabitants of our solar system. For example, the existence of Neptune had been predicted because of its perturbation of the orbit of Uranus. But after the effects of the other known planets were accounted for, a residual precession of Mercury's orbit remained. The excellent book *The Hunt for Vulcan*, published by Thomas Levenson in 2015, recounts the search for the culprit affecting Mercury's orbit.[71] The search ended not when the imagined planet Vulcan was discovered but when Einstein showed that we didn't need it.

As early as 1907, Einstein was seen musing about whether his nascent ideas about a new theory of gravity might eventually be able to explain the orbit of Mercury.[72] And when the calculation agreed, he was triumphant. The final, 1915 version of the theory predicted the orbital precession correctly, and that was immensely important to him.

In a letter to the great physicist and teacher Arnold Sommerfeld in December 1915, Einstein said, "The result of the perihelion motion of Mercury gives me great satisfaction. How helpful to us here is astronomy's pedantic accuracy, which I often used to ridicule secretly!"[73] And the next day, to his friend the engineer Michele Besso, he wrote, "The boldest dreams have now been fulfilled. *General* covariance. Mercury's perihelion motion wonderfully precise."[74]

During this period, Karl Schwarzschild, the scientist who found the first exact solutions to Einstein's gravitational equations, and who is credited with the theoretical discovery of black holes and wormholes, was in correspondence with Einstein about the young theory. He shared Einstein's excitement about the successful Mercury calculation, writing to Einstein in late December 1915: "It is a wonderful thing that the explanation for the Mercury anomaly emerges so convincingly from such an abstract idea." In the same letter, Schwarzschild comments on the raging war: "As you see, the war is kindly disposed toward me, allowing me, despite fierce gunfire at a decidedly terrestrial distance, to take this walk into this your land of ideas."

In the middle of February 1916, Einstein is still writing of the Mercury agreement as centrally important. He wrote to Otto Sternin these telegraphic lines, apparently in a hurry: "General relativity is now, almost exactly since we last saw each other, finally resolved. General covariance of the field equations. Perihelion motion of Mercury explained exactly. Theory very transparent and fine. Lorentz, Ehrenfest, Planck, and Born are convinced adherents, likewise Hilbert."[75]

By early 1917, Einstein seemed to feel that one phase of his work was done: "Scientific life has dozed off, more or less; nothing is going on in my head either. Relativity is complete, in principle, and as for the rest, the slightly modified saying applies: And what he can do he does not want; and what he wants he cannot do."[76] This comment was written in the context of dealing with the depressed wartime economy of Europe and the necessity of carefully hoarding foodstuffs against rationing and shortages. Einstein's general satisfaction is due to his belief that he had resolved the issues of the size of the universe, boundary conditions, and related cosmological problems, at least as far as current astronomical knowledge allowed.

The agreement of Einstein's calculations with the observed orbit of Mercury was a great accomplishment, but the planet's orbit was, after all, something that had already been observed. The real test was to predict things not yet seen. To this end, Einstein made several predictions of phenomena that he hoped astronomers could look for and measure. One of these worked out in such a way as to raise general relativity from an obscure theory understood by a handful of experts into something that was talked about on the front pages of newspapers, and to transform Einstein's life forever.

And because of this, the science and math around general relativity would become a firmly embedded part of physics culture. This overall understanding and acceptance of general relativity would sustain physicists' awareness of the central idea of this book, the theorem proved by Noether and described in the next chapter, so that this theorem would survive long enough to take the central role in the next great chapter of physics.

3

The Theorem

"It would not have harmed the Göttingen old guard to have been sent to Miss Noether for schooling."

In 1918, Emmy Noether published a paper dedicated to Felix Klein "on the occasion of the fiftieth anniversary of his doctorate."[1] Its title can be translated as "Invariant Variational Problems."

It's a modest title that offers no hint of the revelations within its pages. They are difficult pages to follow, even today, and must have been quite a challenge at the time, as the paper makes free use of the language and results at the cutting edge of several areas of mathematics. Even the renowned mathematician Cornelius Lanczos, certainly a modern expert in some of the techniques that Noether brought to bear to attack her problem, had to admit that "the original paper of Noether is not easy reading."[2]

The difficulty is entirely due to the nature of the material and the verbal condensation demanded by the constraints of publication. The exposition itself does not belie Noether's reputation as an elegant and clear expositor. Although she was writing for the small and highly specialized audience of her fellow research mathematicians, even those outside the treated subspecialty can glean here and in her other papers some idea of the context and aims of the work.

In this chapter, we'll first examine the content, meaning, and importance of Noether's theorem: the aforementioned "single most profound result in all of physics," again to quote Frank Wilczek, who received the Nobel Prize in Physics in 2004 for his theoretical work with elementary particles.[3]

I then describe how Noether came to be working on a problem that led to her theorem and how she was uniquely able to make the connections that resulted in its discovery. I'll briefly and qualitatively portray the areas of mathematics that she combined in an original way to extract new knowledge, exposing a hidden connection between two fundamental ideas in physical science. I'll show how her result clarified much of our understanding of the previous several centuries of physics and opened a path for the physics of the future. We'll see how, in proving her theorem, Noether also resolved a crucial issue that remained at the center of the recently published theory of general relativity. And we'll then return briefly to the feverish months leading up to its publication and see again how she became one of its uncredited authors.

Although I can explain the intuitive and physical content of Noether's theorem without using the customary symbology of mathematics, I cannot, without resorting to equations, explain in detail how it is proven. Consequently, you will not find here a detailed recounting of her arguments or methods, for such an explanation lies far beyond the scope of this book. For my omission of the mathematical details, I hope you can forgive me. The demonstration of a mathematical result is a mathematical argument, not translatable to nonmathematical language. This is so, even if the result (fortunately for us) has a strong intuitive content.

Noether's theorem exposes a connection between two ideas that enjoyed a venerable status in physics and that also far transcended, in various guises, physical science: *symmetries* and *conservation laws*. To understand what the theorem says and why it opens up new avenues of thought, we first need to understand these two simple ideas.

The idea of symmetry is timeless. It was often discussed by the ancient Greeks in philosophical and mathematical contexts. It has been with us throughout our intellectual history and has always played a central role in all sciences. Symmetry, of course, exists beyond science, in philosophy and aesthetic theory in particular.

Webster's Revised Unabridged Dictionary, in its superior 1913 incarnation, offers these characteristically elegant definitions of symmetry: "A due proportion of the several parts of a body to each other; adaptation of the form or dimensions of the several parts of a thing to each other; the union and conformity of the members of a work to the whole." The dictionary adds a secondary definition pertaining to biology: "The law of likeness; similarity of structure; regularity in form and arrangement; orderly and similar distribution of parts, such that an animal may be divided into parts which are structurally symmetrical." These are fine general definitions, but later I'll offer one both simpler and more precisely attuned to the way the term is used by mathematicians and physicists.

Symmetry is everywhere in the analysis of painting and music and in psychological explanations of our sensations of beauty in the human form. It plays a descriptive and weakly explanatory role: it is supposed that symmetry pleases us because it brings a kind of logical organization to disparate elements. We are attracted to symmetry in our fellow humans because it's a sign of health and reproductive fitness.

Symmetry is not absent even from moral philosophy, for what is the Golden Rule but an appeal to a principle of symmetry?

In biology, it's a constant source of questions and answers, both in its presence and in its absence. Internally, the disposition of our organs departs from our external bilateral symmetry, but why is the liver on the right instead of on the left? And why, in the small minority of individuals evincing a mirror reflection of the customary internal arrangement, is it so? The shape of a helix defines a type of symmetry, and the unraveling of the helical structure of DNA opened a new epoch in the science of life.[4]

One of the most intriguing mysteries at the intersection of symmetry and biology has to do with the chirality, or handedness, of the amino acids and sugar molecules found in living things. We're all familiar with the fact that, to tighten a screw, we usually have to turn it clockwise: these are right-handed screws. But we also know that we occasionally encounter a left-handed screw; we have to keep an open mind when our repairs aren't going as planned. The reason that almost all screws are right-handed is that we've decided to adopt that as a standard, to make life easier. This right-hand-screw dominance in hardware is a type of asymmetry: in the absence of the chosen convention to make right-handed screws the standard, we would expect to encounter equal numbers of both possibilities. Some of the molecules of life also have a definable chirality, as they have a generally helical form, like a screw; they can be left-handed or right-handed. A departure from symmetry on the molecular level presents an enduring mystery for biologists: Why are all of life's amino acids left-handed, while all the sugars are right-handed?[5] The symmetry of the laws operating at these scales doesn't seem to favor either chirality, so we would expect an indifference where we observe a firm exclusivity.

In physics and astronomy, symmetry has always occupied center stage. The taxonomy of crystal forms finds its explanation in the possible symmetrical arrangements of molecules. The wanderings of the electrons in those molecules, the patterns of vibration in a timpani, and the arrangements of clouds all obey symmetrical solutions of similar equations. Here, too, a conspicuous absence of symmetry occurring with no obvious, compelling explanation creates a mystery: Why do we live in a universe of matter rather than antimatter?

Most of the elementary particles of which we and our stuff are made have an antimatter twin.[6] The twin is identical except that it has the opposite electrical charge. For example, the antimatter twin of the electron is called the positron. It has the same mass and other physical properties of the electron but a positive charge. We can create antimatter particles in accelerators and have even succeeded in putting these particles together to form antimatter atoms, demonstrating, in principle, that antimatter particles can combine in the same way that normal particles combine to create larger structures—structures that presumably form everything we observe in the universe, including ourselves.

Here is where the puzzle arises: the fundamental equations of physics that describe the behavior of elementary particles are indifferent to whether these particles are matter or antimatter. Yet we're made of electrons, protons, and so on, not positrons and antiprotons. Our understanding of the Big Bang is that equal numbers of particles and antiparticles should have been created. Once again, the obstinacy of observed reality in ignoring the symmetrical perfection of our models requires explanation and provokes research. This mysterious violation of symmetry echoes the curious preference of biology for left-handed amino acids and right-handed sugars. The research is ongoing. I describe no answers to these questions, because there are no answers that explain everything or that satisfy everyone.

All of these examples reflect the roles that symmetry considerations played in science and philosophy *before* Noether's theorem: symmetry was a guide to the overall patterns of reality. It served as a concise description and sometimes as an aid to constructing the solution to a problem. Its role was qualitative or semiquantitative. Symmetry told us what to expect and motivated us to explain its absence where we had expected to see it.

Until the arrival of Noether's theorem, symmetry was not law. If it were law, there could be no exceptions. It did not *govern* the way the universe behaved.

Noether's theorem established a new role for symmetry. The ancient companion and guide to humanity's unending project to describe our universe now had new powers. The theorem explained precisely how symmetry *did* govern what could and could not happen wherever it was present. Noether's theorem raised symmetry from a description of a pattern of reality to an active participant in its behavior. It promoted symmetry to the status of law.

The theorem was a signpost along the path of development of critical thought. To understand how deep its influence runs, we need first to grasp, in full generality, the concept of symmetry itself.

The generalized idea of symmetry, the particular conception of symmetry that appears in Noether's theorem, is an extension of the familiar, visualizable examples that immediately come to mind. The word *symmetry* often evokes, and is illustrated by, a butterfly or an idealized face. Both images appeal to our aesthetic instincts; both are symmetrical because one-half is the mirror image of the other. Consequently, the entire object, when admired in a mirror, retains its appearance.

Simple geometrical figures embody either this symmetry or another one (or both): we can rotate them by a certain angle without changing how they appear. A circle can be rotated by any angle; a square, 90 degrees; a general rectangle, 180 degrees; and a hexagon, 60 degrees.

In each case, we have applied a *transformation* (a rotation or a reflection, in these examples) without causing any change. This quality is the key to understanding the generalized concept of symmetry: something has a symmetry if it is the same in every way after you apply a transformation. The type of transformation allowed defines the type of symmetry.

Once you've absorbed this idea, you can take it a step further: The entities embodying symmetry need not be physical objects. Nor do the transformations need to be limited to geometrical operations.

A couple of concrete examples will make this notion, which at first contact might seem forbiddingly abstract, easier to grasp. Imagine that after a period of inflation, a government decides to redenominate its currency.[7] The population trades in their million-unit banknotes for one-unit banknotes; everyone's bank account is divided by a factor of a million. Nothing in the economy changes, except that it's more convenient to talk about money. Checks need fewer zeros. The unit of currency is simply rescaled. It takes the same number of minutes of labor to buy a loaf of bread. The transformation here is the rescaling in the unit of measurement for money; the entity that has remained the same after the transformation is the economic system. The economy is symmetrical with regard to currency rescaling. Various governments have, at times, taken advantage of this symmetry to rescale their currencies after suffering extended bouts of hyperinflation.

Another example: in the United States, we call the lobby floor of a building, the floor that you are on when you enter from the street, the first floor. We count up from number 1 (ignoring the quaint superstition that leads to omitting the number 13). In some other countries (France, for example), the first floor is one floor up from the lobby. The difference can be thought of as where to put an imaginary zeroth floor. In the United States, floor zero is one floor underground, while in France, it's the lobby (the *rez-de-chaussée*). Imagine that one day France decided to come to its senses and adopt the American system. Its apartment buildings and hotels would not suddenly sink into the ground or take flight; nor would they stretch or shrink. The transformation in the location of the zero would have no effect on the reality of the structures themselves. Their heights and positions are symmetrical with regard to the change in convention.

In both these examples, we have a transformation that has no effect on reality: a symmetry, in the wider sense of the term. It is this wider sense that I intend whenever, from this point forward, I use the word *symmetry*.

So much for symmetry. The second big idea is that of a *conservation law*, a rule that demands that a *conserved quantity* does not change its value as time passes. This idea is even simpler than the idea of symmetry. In fact, Noether's theorem derives much of its profundity because it reveals a hidden connection between two ideas that are so simple and that pervade much of our reality and our descriptions of it.

An example of a conservation law is the conservation of mass: the ancient concept of the invariability of the total quantity of substance. You can divide the stuff of reality into smaller bits, and you can stick the bits together, in various combinations, at will. But the total amount of stuff will never change. You cannot bring matter into existence; nor can you banish it from the world. You can even burn a piece of wood, but if you gather all the smoke, ashes, and water vapor, you will have the same mass as the log you started with. Matter can change form, but the *amount* of it is conserved. (Some readers may be aware that the conservation of substance is not strictly true; because of $E = mc^2$, mass and energy can be transformed into each other. But I describe a conservation law as stated and known since antiquity; we can examine its consequences even if it fails to precisely describe the actual world. In any case, we can recover the law by replacing mass by a particular combination of mass and energy.)

Conservation principles can appeal to our intuition as deeply as symmetry. Some people are prey to an inner conviction that their luck is conserved. If something beneficial befalls them, they suffer anxiety while waiting for the other shoe to drop; the next unfortunate accident confirms their belief that every turn of chance must be balanced by one that moves their fortunes in the opposite direction. A similar superstition is known to students of probability theory as the *gambler's fallacy*. It appears as an article of faith among many habitués of casinos. These people believe, for example, that a run of red numbers on a particular roulette wheel somehow makes a subsequent black number more probable than usual. Underlying this belief, which is the basis of innumerable groundless systems of betting, is an unshakable intuition that there must exist some causal force that evens out the results of chance: a law of conservation of probabilities. Of course, the gambler's fallacy and other similar superstitions are examples of purely imaginary conservation laws, but their pervasiveness illustrates how principles of conservation are embedded in our sense of how the world should operate.

My final example is a conservation law resting on firmer ground. Something called *angular momentum* basically tells us how much twisting force we would have to apply to get something to stop spinning. If the spinning thing is composed of various parts, its angular momentum is higher the faster those parts are rotating, the heavier they are, and the farther they are from the center of rotation. Everyone has seen ice skaters spinning around and then spinning faster, as if by magic, as they pull in their arms. You are seeing conservation of angular momentum: as the mass of the arms is pulled in closer to the center, the skater has to spin faster to keep the angular momentum the same. It's not magic; it's just a physical conservation law in action.

The theorem that Noether proved in 1918 revealed a connection between symmetries and conservation laws. But this already understates its significance. For Noether's theorem doesn't merely show how these two ideas are connected; it proves that they are not two separate ideas but are, and have always been, the same idea.

From the earliest records of our long struggle to bring order to the universe, the ideas of symmetry and of conservation have been present in some form. Symmetry appears as an ideal of human creation and as a description of the underlying order of reality. Conservation implies the notions of constancy amid change, of something preserved amid the illusory swirls of contingency.

These two ancient principles evolved into precise statements and took on mathematical forms in the scientific age. They became tools of calculation but never lost their status as profound signifiers of an underlying structure to reality.

Until Noether's theorem, they seemed to embody separate aspects of this underlying structure. With the precision of mathematics, the theorem shows that conservation and symmetry are the same thing.

We need, now, to make precise what "the same thing" means. I've been following a common practice in using the term *Noether's theorem* to refer to what are in fact *four* theorems, all proved in the same 1918 paper. These four theorems are two mathematical statements and their converses. The difference between the two statements is rather technical, having to do with different classes of symmetries; the distinction need not concern us here. But to appreciate the full significance of Noether's theorem, we must be clear about the meaning of *converse*.

A converse of a statement is the statement in reverse. For example, a statement might be "All Greek philosophers are mortal beings." Whether you agree with that or not, let's assume it's true. The converse would be "All mortal beings are Greek philosophers," which I hope you'll agree is not true, even if the original statement is. It's oddly common to encounter people making the mistake of assuming that a statement is the same as its converse, usually in situations where the logical flaw is not so obvious.

Noether's 1918 paper proved that for each symmetry, there was a corresponding conservation law. The paper showed how to derive the conservation law from the symmetry. It also proved the converse: that for any conservation law, there was a corresponding symmetry. This is what I mean when I say that she proved that the two ideas were the same idea: if each implies the other, then one cannot exist without the other. They are, in every sense, the same thing, simply seen in different ways.

The symmetries dealt with in Noether's paper are those of physical systems that admit of a certain description. But this class of systems is quite general and includes, in addition to general relativity (the field that motivated her investigation), any fundamental theory in physics. The class also includes other types of systems that lie beyond physics but can be described with similar mathematical machinery; we'll visit some of these in Chapter 8.

I believe it is this, the existence of these theorems and their converses, that inspires the feeling of awe and profundity in many of the physicists and others who have contemplated them. If the theorems only worked in one direction, they would still be immensely important to physics and its history. But it's the *identity* of symmetries and conservation laws that convinces us that we have reached a new understanding of the beauty and harmony found in nature.

Consider an example. One of the basic and intuitively obvious symmetries in physics is the *time translation symmetry*: the idea that we can place the zero of time anywhere convenient without affecting the predictions that physics makes about how anything will behave, evolve, and change. This idea is so obvious that most people never pause to consider it. Of course if we change, for example, where to put the year zero on our calendars, doing so has no effect on history itself, on what happened, or on how long it took to happen. This irrelevance of the placement of zero time is a symmetry, because it is a transformation that has no effect. It is the same as the previously mentioned numbering of the floors in a building and how the choice of the French or the US convention makes no difference.

Noether's theorem demonstrates something unexpected about the symmetry of time translation: that it is equivalent to energy conservation. And because the theorem includes the converse, it's not that time translation symmetry simply implies energy conservation, or the reverse. It's that time translation symmetry *is* energy conservation.

Energy

After the dust settled at the end of 1915, when Hilbert had relinquished any claim to priority over the general theory and when he and Einstein were friends again, Hilbert turned his attention to an unresolved issue. He had tried several times to show that the general theory of relativity obeyed energy conservation, with no success. By now, scientists accepted that any reasonable theory in physics should obey energy conservation: it was obeyed by the central and fundamental theories of Newton, which explained motion, and those of Maxwell, which covered electricity and magnetism. It was the law. Just as money cannot appear or disappear from your checking account without a deposit or a withdrawal to account for it, energy should not appear or vanish from space unless its movement could be accounted for.

Here I need to be a little more precise about two forms of energy conservation. We can call these forms *global* and *local* energy conservation.

The global type is what almost everyone is familiar with on some level. It's usually stated something like this: in an isolated system, the total amount of energy cannot change over time. When some number—whether it represents a quantity of energy or of something else, calculated from the properties of a system—does not change over time, we call that number a conserved quantity. The statement that the total energy cannot change over time is a statement of energy conservation.

Physics defines various forms of energy. The law of energy conservation means that as an isolated system evolves, some amount of one form of energy can be transformed into another form and sometimes back into the first form, and so on, but the *total* cannot change.

In the energy conservation law, we're careful to stipulate an *isolated* system because if we allow interactions with the world outside the system, all bets are off. In general, these outside interactions—this lack of isolation—will involve some energy transfers into or out of the system, so we wouldn't expect the system's energy to remain unchanged.

As an example of an isolated system, take a collection of billiard balls rolling around on a table and occasionally colliding and changing direction. Of course, in real life, no system is truly isolated. Can you hear the clack of the balls colliding? Those sound waves carry energy out of the system, never to return. But part of the art of physics, and of thinking up idealized examples with which to explain physics, is to focus on the essential and to disregard small departures from an idealized picture.

We try to ignore insignificant effects. The amount of energy carried away by these sound waves is pretty small, a tiny perturbation to the system. Larger is the effect of friction, which we can't ignore and which is a bit more complicated than we'd like to include in our analysis. You know that friction is important because unless, unlike Hilbert, you've avoided pool halls all your life, you know that any particular ball will not keep rolling and bouncing around forever but will pretty quickly slow down and stop.

But we physicists have invented a way to completely eliminate the effects of friction and all other related, complicated phenomena: we simply say we're talking about an *ideal* pool table. In this case and many others, with this one word, *ideal*, we can vanquish all the dragons of heat, friction, dissipation and ignore any of the other forms of energy aside from the energy of motion and other ones introduced a bit later in this chapter.

Using the assumption of an ideal system, we can begin again, this time more carefully. For an example of an isolated system, we can consider a collection of billiard balls rolling around on an ideal table and occasionally colliding and changing direction. We neglect the effects of friction and other phenomena. Here we have only one form of energy to keep track of: the kinetic energy of the system, which is the total of the kinetic energy of all the individual balls.

Kinetic energy is the modern term for the energy of motion, which we can calculate from an object's mass and speed. This form of energy is proportional to its mass and to the *square* of its speed, explaining why it's so much worse to crash your car into a tree at forty miles per hour than at twenty miles per hour.

In the ideal case, where no energy is lost to sound, heat, or any other factor outside our ideal model, if we know the system's kinetic energy at any particular time, we know it for all times before or after that, for it can never change. That's because energy conservation is the law, and kinetic energy is the only kind of energy in this system.

Knowing about energy conservation on the billiard table makes it far easier to solve certain types of problems, especially because all the balls (in the pool or billiard games with which I'm familiar) have the same mass, so we can just focus on the velocities. For instance, suppose I place two balls close to each other, or touching, and shoot a third ball at the spot exactly between them at a known speed. I observe that the third ball comes to a stop as the first two each fly off, and I wonder if I can calculate what the speeds of the originally stationary balls have to be.

Using a detailed force calculation and Newton's laws of motion would be one way, but energy conservation makes it so much easier. We know from the geometrical symmetry of the problem that the first two balls must end up with equal speeds. And from the law of energy conservation, we know that the energy on the table after the collision must be the same as the energy that I imparted to the third ball. (Obviously during my interaction with the balls, the system is not isolated, but after I hit the ball and step back, the system evolves in isolation, at least according to the rules of the game.) Using just this law and knowing the expression for kinetic energy, I can calculate in my head that the speed of each of the first two balls will be the speed of the third divided by the square root of two (I certainly hope that's correct). Conservation laws are a powerful technology for solving problems.

One more example will show how energy conservation works when we have more than one kind of energy. If there are springs, electricity, gravity, or other sources of force (other than instantaneous forces such as from ideal billiard ball collisions), then the total *kinetic* energy of the system is *not* conserved. We need to add to the energy of motion something called *potential energy*, which represents stored kinetic energy that is available to be returned to the masses as motion again.

With actual springs, we can see this potential energy in their compression or elongation away from their equilibrium positions. But with gravity or electricity, the idea is a bit more abstract. We see the changes of potential energy in the *configuration* of the system, that is, in the positions of its constituent parts.

One example is a cannonball shot straight up from the ground. Initially, the ball has a very large quantity of kinetic energy, but as it rises, the earth's gravity slows it down until it comes to rest, for an instant, at its zenith. As it slows on its way to its zenith, the cannonball's kinetic energy decreases, but the energy of the total combined cannonball-earth system remains the same, because the ball's energy is stored in the form of potential energy. This potential energy is at a maximum when the ball comes to rest at its maximum height and the kinetic energy is zero. Immediately, the potential energy begins to be converted back into kinetic energy as the cannonball picks up speed on the way down. When the object reaches the ground, its kinetic energy, and therefore velocity, is the same as when it first left the barrel. (All these observations neglect the friction of the atmosphere and other sources of energy dissipation.)

As in the case of the billiard balls, it's almost always easier to solve cannonball problems by exploiting energy conservation than by calculating forces and solving the differential equation equivalent to Newton's second law of motion.

Energy conservation may be so familiar to us that we assume it is somehow obvious. But Newton did not know about it, and the law was not discovered, in its definitive form, for almost two hundred years after he published his laws of motion.

Since energy conservation was not yet part of science, it was not part of the wider culture, either. Shakespeare was generally aware of the scientific developments of his time.[8] If the seventeenth-century worldview had included the necessity of the conservation of energy, he surely would not have written, in his Sonnet 154, "Love's fire heats water, water cools not love," for he would have understood that if one substance supplies energy to a second substance, then the first must lose the same quantity of energy that the second gains.[9]

By 1915, energy conservation had been accepted as a general law of the universe. At least, no exceptions were known.

In addition to simple classical systems such as the ones in the preceding examples, energy (and momentum) conservation worked with more complex field theories (see the previous chapter) such as electromagnetism and hydrodynamics. Energy conservation in field theories takes the form of a *local energy conservation* law rather than the simpler global energy conservation just described. Of course, global energy conservation still holds as well and can be derived from the local law, but the local law must be satisfied as well.

In cases such as electrodynamics, we still need a way to express the idea that energy is neither created nor destroyed. We must do so even when energy cannot be neatly assigned to individual objects, like billiard balls, flying around our system, or easily recognized as resulting from the position of a cannonball.

Local conservation laws are stricter than global ones. They require not only that the total amount of energy remain constant in time but that, in any region of space within the system, the change in energy be accounted for by the flow of energy into or out of the region. In other words, local conservation laws formalize the idea that energy cannot be created or destroyed. And this requirement applies not merely to the total, in an entire system, but to any region of the system, whether large or microscopically small.

The fact that general relativity didn't seem to obey a local energy conservation principle was the problem that nagged at Hilbert. As far as he and Klein could see, the new theory of gravity did not respect this bedrock principle of nature.[10] Apparently, energy could be illegally created and destroyed.

They naturally considered this a fatal flaw, one so profound that they wondered if they had made an error in calculation or were perhaps suffering some deep conceptual confusion. But as Hilbert repeated the calculations and as his colleagues tried to find ways around it, the problem seemed unavoidable. The situation could lead to paradoxes. When the possibility of gravitational waves was considered, it was found that an object could speed up as it lost energy by emitting gravity waves, rather than slowing down, as one would expect.

The Göttingen mathematicians were willing to accept a theory radically different from anything that had been even suggested before as a model of material reality, and the theory of general relativity certainly fit this description. Einstein's idea turned the very structure of space and time into a player on the dynamical stage, subject to continuous change, just like the planets and other masses moving through what used to be considered the unchanging background of space. They could accept all this, as could at least part of the small fraction of the physics community capable of understanding it. But the failure of energy conservation would be a deal-breaker, at least according to what had become the received wisdom in the physical sciences. Unless this apparent defect was resolved, the theory of general relativity might be seen as a fascinating mathematical curiosity but could never be seriously considered as a description of the real world.

Hilbert was not the first person to worry about energy conservation in general relativity. Einstein himself had returned to it periodically during the development of the theory over the previous eight years or so. He had found no satisfactory solution that worked with the final, correct version of the gravitational equations.

Klein had also been communicating with Carl Runge, a well-respected mathematical physicist, about the energy issue in general relativity. Those who have studied elementary numerical methods have encountered the Runge-Kutta method; this is the same Runge. His ability to delve so deeply into Einstein's theory so soon after its dissemination certainly indicates a high level of mathematical sophistication.

Runge thought that he had found a way to get general relativity to obey the local conservation law.[11] Impressed by Runge's concept, Klein showed it to Noether. But she immediately saw that Runge's idea wouldn't work, and gave Klein the bad news.

The situation in 1916, therefore, was that the status of energy conservation in general relativity was confused and uncertain.

───────

Hilbert and Klein had been keen on getting Noether to Göttingen partly because of her expertise in invariant theory, especially aspects of it that still interested the two men. Hilbert, of course, was already deeply curious about what Einstein was doing with gravity in 1914, having tried to coax him to pay a visit to Göttingen starting in 1912. It's unclear whether Hilbert had foreseen in particular that he himself would be studying general relativity as a practitioner and, further, that he would benefit from collaborating with Noether. It seems unlikely. There's no sign that Hilbert put his pen to paper to work on the Einstein version of gravitation before the physicist finally came to give his lectures. As we've seen from Noether's letters, Hilbert had started working on general relativity immediately after those lectures and had, from the start, brought its many mathematical puzzles to her. Once he hit the wall of energy conservation, it was clear to him that some of the methods of invariant theory might be brought to bear on the problem. He asked Noether to see if she could break through the wall.

Although Noether had no particular interest in physics, she had done her homework. She would develop a reputation in later years as a thorough scholar, intimately familiar with the literature, both historical and recent, in every field of mathematics that she explored. Her 1918 paper shows that she was keenly aware of the physical implications of its content; she knew that it had relevance both to science and to pure mathematics. She refers to its connection with general relativity throughout and mentions several special cases of her general theorems that had been unearthed in mechanics in the historical literature. She had no way of knowing that physicists and mathematicians, less conscientious in their scholarship than she, would "discover" special cases of her theorem in the future, down to the present day, and believe each time that they had found something new.[12]

We're now in possession of a more general idea of symmetry: the quality that remains unchanged in an object after we apply a transformation. Similarly, a physics theory has a symmetry if all its defining equations, or their consequences, are unchanged after you apply the transformation. Take Newton's mechanics. It consists of a few concepts and laws of motion. The second law amounts to a differential equation that we solve to get the motion of a cannonball or anything else. It contains a variable for time, a variable for each spatial dimension, and variables for the masses of the objects. It also includes the forces at play, be they gravity or anything else. The law relates these forces to the *change* in the velocities of the objects, where these velocities are the *changes* in the objects' positions in time.

The variable of time, for example, appears in the second law only in an expression that refers to its change. The "naked" time variable does not appear. This means that we can add any constant value to the time variable or—doing the same thing—change where we define the "zero" of time, and no predictions of the second law or of Newtonian mechanics will be affected. The shift in time just drops out of the equation. In other words, if it takes you an hour to drive to work today, it will still take you an hour to drive to work after daylight saving time goes into effect and we've subtracted an hour from the arbitrary numbers that we use to say what time it is.

The preceding discussion means that classical mechanics is symmetric under time translation (*translation* is the word physicists like to use for shifting some variable). And these continuous time translations do define the type of symmetry structure to which Noether's theorem applies. In the case of time translation symmetry, Noether's theorem tells us that the associated conserved quantity is the total energy. As a physical system evolves, its various types of energy may each be converted into each other, but the sum of them all, the total energy, will not change with time.

Why is this a big deal? We already knew about the conservation of energy, after all. As mentioned, we "knew" about this law of nature but not about why it was true in general. The conservation of energy had a semi-empirical status. It could be proven for some systems. For example, with the cannonball shot, we can prove that the sum of the kinetic and potential energies remains constant. For other systems, scientists just assumed that energy conservation must hold because no exceptions had yet been found. That's why it was expected to hold in general relativity as well.

Noether's theorem showed that energy was conserved in any system with time translation symmetry. This result raised the murky idea of energy conservation to a fundamental aspect of reality, a necessary property of any universe whose basic symmetries agreed with our deepest intuitions. Because all known physical theories respected time translation symmetry, Noether's theorem demonstrated that energy conservation was indeed a universal law. And it elevated the law from a semi-empirical observation to a mathematical truth.

Another example of a simple application of Noether's theorem is the symmetry of spatial translations. Just as the arbitrary location where we place the zero of time should make no difference, the zero of any spatial coordinate should not matter, either. Naked spatial coordinates do not appear in the equations of motion. Another way to say this is that Newtonian mechanics is homogeneous in space. Turning the crank of Noether's theorem shows that the symmetry of spatial homogeneity is equivalent to momentum conservation.

The conservation of momentum, a conservation law that was known before energy conservation, is implicit in Newton's laws. It's the conservation law that explains why rockets work and why you feel a recoil in your hand or shoulder when using firearms. The momentum of the expelled fuel from a rocket or a projectile from a firearm must be balanced by an equal and opposite momentum. A simple version of this idea is expressed in the most familiar of the laws of Newton: for every action, there is an equal and opposite reaction.

But Noether's results showed that momentum conservation was more than a special property of Newtonian mechanics. Instead, it was a general, and necessary, feature of any spatially homogeneous theory. A theory that did *not* have this symmetry would imply that its results—the laws of nature—were somehow different in different places. Since this is an affront to our intuition, we naturally expect that any reasonable physics respect the symmetry of spatial homogeneity. The symmetry implications of momentum conservation exposed by Noether's theorem elevated the conservation law to something that we would expect any theory of physics to encompass. And, as with energy conservation, momentum conservation would be a property of any universe that conformed to our deepest intuitions about the symmetries of space.

The last example of Noether's theory of symmetry concerns the symmetry of rotations, the symmetry inherent in a circle. Noether's procedure showed that this form of symmetry implied the conservation of angular momentum.

Rotational symmetry is also called *spatial isotropy*. A theory that did not respect this symmetry would admit the idea of experiments that gave a different result when repeated after rotating the laboratory. A different result doesn't seem reasonable, so angular momentum, like the other conservation laws, could now be seen as a consequence of intuitive space-time symmetries that any rational theory would have to respect.

Note that these symmetries are symmetries inherent in the *theories*, as reflected in the symmetries of their equations. Clearly, if you perform an experiment twice in different locations where different conditions affect the outcome, that doesn't mean the laws of nature have changed. Implicit in this discussion is the idea that, for example, translation in space means *only* shifting the location in space while holding any relevant influences constant.

Time translation, spatial translation, and rotational symmetry are just three examples of symmetries that happen to be connected, through Noether's theorem, with familiar conservation laws. The theorem proves that any symmetry leads to a conservation law, even if it might not have a name or might be unknown. Presumably, the theorem could be used to discover new conservation laws (and was, as we'll see in Chapter 7).

Because Noether also proved the converse of her theorems, we know not only that a specific symmetry implies a specific conservation law but also that the conservation law implies the symmetry. If you know you have a conserved quantity, then you also know that there is a symmetry in the theory, perhaps hidden, perhaps unknown. But it *must* be there.

As I'll relate in Chapter 7, Noether's theorem was slow to penetrate the consciousness of most scientists. One anecdote demonstrates something of this gradual percolation of the awareness of the theorem. In an interview, Werner Heisenberg remembered how, long before, a paper had attempted to resolve an experimental puzzle in particle physics. The data had suggested that energy was not being conserved. The paper's authors had proposed that energy might be conserved only "statistically," and not exactly. When Wolfgang Pauli (one of the most prominent theorists of the time) heard of this, he said, "Well, that's too dangerous. There you try something which one shouldn't try."

Heisenberg goes on: "Much later, of course, the physicists recognized that the conservation laws and the symmetry properties were the same. And therefore, if you touch the energy conservation, then it means that you touch the translation in time. And that, of course, nobody would have dared to touch. But at that time, this connection was not so clear. Well, it was apparently clear to Noether, but not for the average physicist."[13]

———

We saw in the previous chapter how, in special relativity, space and time, which had separate identities in Newton's physics, were combined into a unified mathematical object, a four-dimensional space-time. In physics before Einstein, energy and momentum had maintained their separate identities. Each was conserved independently of the other, and Noether showed how the separate conservation laws were equivalent to separate symmetries: energy conservation with time translation and momentum conservation with space translation. But in special relativity, since space and time no longer had separate identities, Noether's theorem suggests that energy and momentum might not have separate identities, either.

This is indeed the case. In special relativity, energy and momentum were combined into an object called the *energy-momentum four vector*. (It has this ungainly name because it has four components: one for the energy and three from the three components of momentum in three-dimensional space.) Therefore, the new mathematical object appearing in Noether's paper to represent the conserved entity in general relativity was not of a completely unfamiliar kind to physicists, who had already gotten used to combining energy and momentum. In fact, when special relativity is applied to continuous fields such as electromagnetism or fluids, a related but simpler tensor already appears.[14]

The pre-Einstein energy and momentum conservation laws, which were two different, independent facts of nature, had been combined into one law for the conservation of relativistic energy-momentum. You can see that this had to happen, if you think about $E = mc^2$. Mass was now another form of energy all by itself. It's actually possible for two photons (quantum particles of light), which have no mass, to collide and become transformed into two particles with mass, an electron and a positron.[15] We can't explain the process or predict its outcome using the separate energy and momentum conservation laws.

In general relativity, the situation is considerably more complex. We can still think about something like an energy-momentum four vector, a cousin of the one that appears in special relativity. However, Einstein's earlier theory did not demand a wholesale replacement of Newton's law of gravity or of Euclidean geometry. In general relativity, the gravitational "force" is replaced by the properties of a curved space-time. The curvature of space-time is determined by the mass and energy-momentum contained within it. And here's where it gets really complicated: the curvature causes things to move, which in turn changes the curvature. Also, the curvature of space-time *itself* contains its own contribution to the energy-momentum. This intricate interdependence is what makes the gravitational equations so hard to solve. The energy-momentum and the space-time curvature exist in a nonlinear relationship, each creating the other. In the pithy formulation of physicist and gravity expert John Wheeler, "Space-time tells matter how to move; matter tells space-time how to curve." It took someone with the mathematical talents of an Emmy Noether to untangle this web.

Noether had broken through Hilbert's wall. She was able to show Klein and Hilbert that they were not as confused as they might have thought. They hadn't been able to find a local energy law, because there *could not be one.*

By proving its exact equivalence with the symmetry of time translation, Noether's theorem serves as the *definition* of energy, a concept that, until her result, tended to attract ad hoc and imprecise descriptions.

Feza Gürsey was a Turkish mathematician and physicist who carried out fundamental research in mathematical physics on the applications of symmetry to quantum theory and general relativity.[16] According to Gürsey, "Before Noether's theorem the principle of conservation of energy was shrouded in mystery. . . . Noether's simple and profound mathematical formulation did much to demystify physics."[17]

Often when new entities were introduced into physics, some form of energy was defined in terms of the new entity. It was not always obvious what form this energy should take, but it needed to participate in the inviolable law of conservation and to connect correctly with other physics concepts such as force and work. So, for example, in his monograph laying out his new synthesis of electricity and magnetism, James Maxwell devoted several pages to working out just what form the energy of the electromagnetic field should take.[18]

It wasn't always clear how to define the energy, because, before Noether, it wasn't clear what exactly energy was. In the physics of today, the concept of energy is *supplied* by Noether's theorem. If a physical system possesses time translation symmetry, then we say that it has an energy, and we derive the form of the energy by applying the theorem: the energy in the system is the quantity that satisfies the conservation law that Noether's theorem shows is equivalent to the symmetry. Noether's theorem eliminates the confusion and the ambiguity. If there is no time translation symmetry, as in general relativity, then we know that no energy can be defined.

Einstein's special theory of relativity contains all kinds of bizarre and wonderful phenomena, such as the slowing down of clocks and the contraction of space; these things don't exist in Newtonian mechanics. However, we know that Newton's physics still works now just as well as it did centuries ago. Reality doesn't start to suddenly behave differently because somebody invents a new physics. The resolution to this conflict is that the predictions of special relativity match those of Newton's under a *limiting condition*: that all velocities are much smaller than the speed of light. Nobody had observed time dilation or any of the other non-Newtonian effects, because they were too small to measure. We can measure them now, of course, and they agree wonderfully with Einstein's version of reality.

In the general theory, the relevant limiting condition is that of weak gravity. If we're in a region far from any enormous concentration of mass or energy, the gravity will be pretty weak and space-time will be approximately flat. The geometry of the world will be very close to the Euclidean version that we learn about in high school. The angles of a triangle will add up to 180 degrees. The departure from an elliptical orbit is only observed easily with Mercury because, being the closest planet to the sun, it experiences the strongest gravity—the most warped space-time.

In finite regions of flat, or nearly flat, space-time, energy will be approximately conserved, just as it is in Newton's universe. Since most of space-time is approximately flat, we can calculate many things as if a local energy conservation law holds, even if the equations show that, in general, it does not.

This fact was used in the first detection of gravitational waves, one of the predictions of general relativity (and one with its own stormy history). Similarly to how accelerating charges emit electromagnetic waves, accelerating masses emit waves of space-time distortions that we call *gravitational waves*. These are actual vibrations in the geometry of space-time that travel through the universe at the speed of light. You may remember the exciting news about the detection of gravitational waves in 2015 from the LIGO experiment.[19] The experiment detected them by directly measuring space-time distortions, which caused the length of an object to oscillate. The precision of the measurements required to detect this tiny distortion in space-time is astounding. The results were widely reported in newspapers. For example, a *New York Times* article announced, "Gravitational Waves Detected, Confirming Einstein's Theory."[20] Unfortunately, most headlines and articles claimed or suggested that the experiment constituted the first detection, but it was not. The finding *was* a historic achievement because it was the first *direct* detection of the waves.

However, gravitational waves were in fact first detected in 1974, when two astronomers, Russell Hulse and Joseph Taylor, discovered a pair of (probably) neutron stars zipping around each other in a fast orbit.[21] The binary system is now called the Hulse-Taylor pulsar in their honor. The astronomers could detect that the orbit is slowly decaying, and successfully explained that the decay was due to the energy lost in the form of gravitational waves. The rate of decay agreed exactly with the predictions of general relativity. Hulse and Taylor received the

Nobel Prize for this work about twenty years later. (The aforementioned *New York Times* article is a rare example of a notice in the popular press that did mention the pulsar work.)

The calculation and the verification of gravitational waves from the pulsar depended entirely on energy conservation: the energy loss in the orbit was seen to be exactly balanced by the energy carried away by the waves. But certainly, gravity and the warping of space-time were strong near the neutron stars; if they weren't, there would be no appreciable gravitational radiation to begin with. How can we use energy conservation to verify the predictions of a theory where there is no energy conservation, and specifically in a regime where we know that the conservation law does not even hold approximately?

To understand the answer, we must first appreciate that space is vast, with enormous distances between objects. Near the orbiting neutron stars, space-time will be strongly warped, and we'll need general relativity to describe reality. But let's back away hundreds of light-years from these two massive objects until we reach a region far from any large masses, where gravity is weak and space-time is flat. Now let's draw an enormous, imaginary sphere with its center near the pulsars and its perimeter at our location. Although the sphere is huge, it only contains the pulsars; this is possible because of the vastness of space.

Now keep track of all the gravitational radiation crossing the boundary of the sphere. It carries energy with it, out of the volume inside the sphere. Although in general, as Noether showed, we don't have local energy conservation in general relativity, we have a close approximation of it in regions of flat space. Consequently, the energy crossing the spherical boundary must be balanced by the energy lost from the binary pulsar system.

Of course, nobody created any giant spheres. But the argument shows that if we can calculate the energy carried away by the gravitational waves, we can calculate the consequent decay of the orbit—and that's how the reality of gravitational waves was first demonstrated.

(The debates over the very existence of gravitational waves lasted for decades, involving, among other incidents, Einstein's petulant reaction to having one of his papers rejected by a referee. One productive step was taken by Richard Feynman in 1957.[22] Using a simple thought experiment involving beads sliding on a rod, he convinced a crowd of physicists that these waves could exist and must carry energy.)

———

Noether's theorem can make the analysis of a tricky problem straightforward. Feynman, in his (in)famous *Lectures on Physics*, poses a problem that he describes as a "paradox" (while making it clear that its paradoxical nature is merely apparent).[23]

In Feynman's thought experiment, a metal disk, part of a simple electrical apparatus, suddenly begins to rotate, even though no obvious turning force acts on it. It's as though the ice skater, in our earlier example, began to spin faster without any movement of the arms. In either case, angular momentum appears out of nowhere. Its conservation seems to be violated.

The resolution of the paradox is that, of course, angular momentum is still in fact conserved, as it must be. It was zero at the start, when nothing was moving, and is still zero after the disk has begun to rotate. Since the angular momentum remains zero, never changing, the conservation law is not violated. But how can this be, when the disk clearly turns and therefore obviously attains some nonzero angular momentum? The resolution is that the angular momentum of the disk is canceled by the equal and opposite angular momentum *in the electric field* surrounding the disk. A positive and a negative add to maintain the initially zero angular momentum. Feynman warns that this is a difficult problem, presumably because the student may not realize that the invisible force fields called electromagnetic fields can themselves carry angular momentum.

However, the student who is aware of Noether's theorem will not only understand that electromagnetic fields have angular momentum but will also be able to use the theorem to calculate it exactly. This is another instance where the theorem not only confirms a conservation law (which holds in this case because of the symmetry of spatial isotropy) but also identifies the form that the conserved quantity takes. Here Noether's theorem turns a conceptually confusing problem into an almost mechanical calculation.

─────

Through its clarification of the status of conservation laws in general relativity, Noether's theorem has important implications for *cosmology*, the study of the origin and evolution of the universe as a whole.

The idea of the law of conservation of energy is drilled into every physics student, usually without the context provided by Noether's theorem. This omission is mainly due to the habit of putting off more general, symmetry-based approaches to physics until advanced classes typically at the graduate level. In advanced classical mechanics, the student may encounter Noether-style approaches, always in a watered-down form and often with no mention of Noether herself.[24] Physicists who become exposed to high-energy physics necessarily learn about applications of Noether's theorem in much greater detail. As explained in Chapter 7, the entire theoretical framework of this field is based on the theorem or special cases of it. Most of a student's undergraduate education, and much of the coursework in graduate school, therefore pursues physics in something closer to the Newtonian style.

The result of this suboptimal educational tradition is that unless they specialize in cosmology or gravitation, many grown-up physicists believe that the conservation of energy is an inviolable law of the universe. They are surprised at the assertion that the energy of the universe as a whole is not conserved, often insisting that the assertion must be incorrect.[25]

The same physicists accept that the universe is expanding. Quantum mechanics can provide some insight into the consequences of that.

A photon of light has an energy inversely proportional to its wavelength. Long-wavelength photons, such as those from a microwave oven, have less energy than shorter-wavelength photons, such as those from an x-ray machine. That's why ultraviolet light (short wavelength) from the sun can cause cancer, but cell phone radiation (long wavelength) cannot. The individual photons must have sufficient energy to damage the DNA molecules inside our cells. Saying that radio waves from a cell phone can cause cancer is similar to saying that you can knock over an elephant with a barrage of soap bubbles.

In an expanding universe consisting mainly of light, an observer sitting anywhere will observe all of its photons receding—their wavelengths will be shifted to the red (longer-wavelength) side of the visible spectrum—*redshifted*—just as the light from distant stars is redshifted in our actual universe, because of the Doppler effect. The Doppler effect is the same mechanism that causes a siren to have a higher pitch as an ambulance approaches you, only to shift to a lower tone as it moves away. The effect works for any wave phenomenon—light as well as sound. Since light shifted to the red has a lower energy, the total energy of this hypothetical universe must continuously decrease as it expands.

Noether's theorem lets us prove that the total energy in an expanding universe decreases, as a direct consequence of general relativity, in agreement with the argument from photon energy. It resolves what might otherwise seem, at least to some, to be a paradox of cosmology.

In a more general sense, however, it is in fact impossible to define a quantity to fulfill the role of energy for the entire cosmos. We now know that Noether's theorem clarifies the meaning of energy by identifying it with the conserved quantity that is equivalent with time translation symmetry. This approach replaced the older, ad hoc notions of energy and is what modern physicists turn to when a precise definition for energy is needed. But as we also now know, if the cosmos is described by general relativity, then it lacks the symmetry of time translation invariance, unlike the more intuitive Newtonian universe. If the cosmos is expanding or contracting, then the conditions are clearly not constant in time. Unlike the Newtonian universe, in Einstein's universe, our definition of time zero makes a difference. Today's same experiment cannot be repeated tomorrow, even in principle, because by tomorrow, today's universe will no longer exist. If we don't have time translation invariance, then we don't have an energy definable through Noether's theorem. We are led to the conclusion, still uncomfortable for some, that in our universe, no sensible total energy can be defined.[26] This is even more obvious in some hypothesized universes, in which a period of expansion is followed by a contraction stage and in which we will eventually return to Big Bang–type conditions. As we approach conditions similar to the Big Bang, space-time curvature will again be extreme and there will be no region of flat space-time. In that scenario, there can be no quantity that behaves even approximately like an energy over the entire past and future history of that universe. The situation is perhaps best summed up in a widely used textbook on Einstein's theory of gravitation: "The issue of energy in General Relativity is a rather delicate one."[27]

The question of energy in general relativity is still a matter of active research. The research has expanded since 1918 in step with the expansion of our knowledge of the universe, which has grown enormously. It's easy to forget that when Noether proved her theorem, most people thought that our Milky Way galaxy was all there was. What's more, they presumed that the universe was *static*, eternally unchanging in size. In fact, when Einstein realized that his equations permitted solutions where the universe could expand or contract, he considered this result a defect and adjusted the theory to eliminate those clearly unreasonable possibilities.[28] Now, beginning with Edwin Hubble's discoveries that distant objects were receding from us, we've lived with an expanding universe for so long that the notion seems quite natural.

Noether's theorem is a guide that helps us make sense of our cosmos, whether it will expand forever or one day collapse in a Big Crunch. By relating space-time symmetries to local and global conservation laws, the theorem delineates what is possible and shepherds our intuition into uncharted territories.

———

Emmy Noether found a solution to the energy conservation problem that had perplexed the great David Hilbert, stumped his other colleagues, and eluded Einstein. She did so by proving a set of related theorems, mathematical results of great power and generality. Although nobody realized it then, her result would eventually transcend the original context. It would profoundly change all of physics and, a hundred years later, begin to transform fields ranging from biology to economics.

We ultimately can't explain how a mathematical genius manages to discover profound connections between things that have eluded everyone else. But in analyzing how her theorem arose from the contact between at least two widely separated areas of mathematical research that were brought together by a rare mind working at the forefront of each of them, we've taken a step toward an intuitive appreciation of at least one aspect of the creative process.

As briefly described in Chapter 8, one unexpected application of Noether's theorem in recent decades has been in economics. Some of the economists who took the lead in this innovative exploitation of Noether's mathematics also spent some time musing about how she was able to make such a breakthrough. "Noether had the ingenious insight of combining the methods of the formal calculus of variation[s] with those of Lie group theory," they note, referring to an advanced form of calculus and part of the formal mathematics of symmetry.[29] Many observers of important acts of creation in science, the arts, and technology have noticed a recurring pattern: far-reaching results often arise at the intersection of fields that have developed separately until they are brought together by an unusual individual who is intimate with all of them.

This was the case with Noether. By the time Hilbert had placed the problem of energy conservation in gravitation before her, she had become one of the world's supreme experts in two quite different fields of mathematics. One field was related to the theory of invariants, as described in Chapter 1. This was the mathematics of symmetry in its most general form. Her greatest fame among mathematicians would eventually result from her refinement and rebuilding of related areas of mathematics, which became what we know today as *abstract algebra.*

The second field is the aforementioned *calculus of variations* (or variational calculus) and a nexus of related theories. The calculus of variations, an evolution of the calculus invented by Newton and Gottfried Wilhelm Leibniz, can calculate entire histories or paths of systems. For example, calculus can give us formulas for the speed of a baseball at any moment, but the calculus of variations can tell us its entire path as the solution to a single problem.

Perhaps it's not too far afield at this juncture to remember what Alan Kay, the creative computer scientist and inventor of the Smalltalk programming language, said about this subject: "All creativity is an extended form of a joke. Most creativity is a transition from one context into another where things are more surprising."[30]

As I mentioned earlier, Hilbert's personal formulation of general relativity was, in a sense, more concise and elegant than Einstein's version. Hilbert had found a way to write the gravitational field equations as a compact statement in the language of the variational calculus. But most histories of this episode fail to point out that he depended on some of Noether's results to do so.

Noether's Contributions to General Relativity

The "race" between Einstein and Hilbert to reach the correct gravitational field equations, recounted in the previous chapter, is usually depicted as two men scribbling mathematics in their respective solitudes and happening to reach the finish line through quite different routes. More detailed histories mention that Hilbert was happy to share his progress with Einstein, in whose fevered brain the somewhat paranoid idea of a competition germinated and festered.

There is another inaccuracy in the way the story is usually told. Although sparse, the totality of the correspondence, notes, and remembrances from this brief period makes it clear that Hilbert was not working alone. He was conferring to an extent with Felix Klein, but Klein's own letters testify to the fact that he was more or less out of his depth. The only colleague at Göttingen, aside from Hilbert, who we are certain had complete fluency in the mathematics and methods requisite for attacking every aspect of general relativity was Noether. As historian of mathematics David Rowe points out, all of Hilbert's and Klein's work relating to general relativity "relied heavily on Emmy Noether's expertise in differential invariant theory."[31]

Apparently, far in advance of Noether's formal publication of her theorem, she had worked out much of the results and shared them with Hilbert. We also know that Hilbert's version of the gravitational field equations depended on a theorem that he discovered and that this theorem in turn depended on a version of Noether's theorem that he was aware of before November 1915. The unavoidable implication is that Hilbert's formulation of general relativity, which, as mentioned, employed the language of the variational calculus and was developed independently of Einstein, is the joint work of Hilbert and Noether.

Questions of credit and priority were not of any particular interest to Hilbert, who was focused on results—on seeing mathematics make progress. His attitude in this was shared completely by Noether. But perhaps his knowledge that "his" version of general relativity was not his alone contributed to his readiness to resolve the looming priority dispute with Einstein by abandoning all claims once and forever.

By this time, after all that has been written over the years about the genesis of general relativity, it should be obvious that meticulous questions of priority will never be settled. Nor do they need to be settled.

Ideas are like soap bubbles, glistening with the colors of the spectrum and gliding soundlessly from brain to brain. Emmy Noether, David Hilbert, and Albert Einstein are three of the most ingenious minds that have emerged in the entire history of the human species. That they were all three together, asking the same questions at the same time and place, is a miracle of fortune. The theory of gravitation that emerged is a new way to describe reality itself, a sharp turn in the history of human thought.

Calling general relativity a new way to describe reality is not an exaggeration. The other physical theories deal with particular interactions between particular types of entities. Two electrons interact through electromagnetism but not through the strong nuclear force. Two neutrons interact through the strong force, but there is no electrical force between them. But *everything*—even light—interacts through gravity. General relativity has been verified from microscopic distances up to the size of the entire cosmos. It describes space-time itself, and everything—every bit of matter and energy—obeys its laws. It is literally the fabric of physical reality.

And underneath gravity, underneath the quantum theories of electromagnetism and nuclear forces, underneath any future theory yet unimagined, sits Noether's theorem.

Hilbert's interest in general relativity was part of an ambitious but ultimately unsuccessful effort to construct a unified theory that would encompass both gravity and the nature of matter. It was also an expression of his conviction that formal methods, with which he had had such stunning success in mathematics, were the key to putting physics on surer ground.

Einstein's opinion of Hilbert's attempt at a unified theory can be seen in his letter to Hermann Weyl near the end of November 1916: "Hilbert's assumption about matter appears childish to me, in the sense of a child who does not know any of the tricks of the world outside. . . . Mixing the solid considerations originating from the relativity postulate with such bold, unfounded hypotheses about the structure of the electron or matter cannot be sanctioned. I gladly admit that the search for a *suitable* hypothesis . . . for the structural makeup of the electron, is one of the most important tasks of theory today. The 'axiomatic method' can be of little use here, though."[32]

Today it would seem as if Einstein was entirely correct. Despite the unparalleled importance of Hilbert's program for pure mathematics, a highly formalized approach to theoretical physics has rarely been useful in the discovery of new science. Physicists tend to employ a more Babylonian style of mathematics, in contrast to a classical Greek style, to use the nomenclature introduced by Feynman in a fascinating lecture (available online).[33] In his talk, Feynman explains what he means by his terminology: physicists tend to use a loosely connected bundle of mathematical results, selecting formulas as they find them useful. While having some idea of how to get from one formula to another, they do not work within a formally defined axiomatic system, with all the definitions and assumptions carefully spelled out and all the derivations perfectly rigorous. This more systematic style of mathematics is what Feynman associates with the Greeks. It finds its exemplar in Euclid's *Elements*, the timeless exposition and gathering of Greek geometrical knowledge.

Of course, the mathematics that Hilbert discovered (or invented, depending on your particular philosophy of mathematics) *was* immensely useful for all branches of physics. And it will continue to be, as long as mathematics remains the language of science. From Hilbert space to integral equations, theoretical physics would look very different without Hilbert's contributions to the machinery. And his deep mathematical insights did often have clear implications for physics—and not only in general relativity.

For these and other reasons, Hilbert is sometimes painted as the last significant universalist by historians of mathematics. One historian describes him as having led mathematics out of the nineteenth century and into the twentieth.[34]

Other Symmetries

Noether's theorem works with other symmetries besides those that have an obvious physical interpretation, such as the time translation invariance and the other symmetries just discussed.

To understand the full significance of the theorem, we need to appreciate that it applies to other types of symmetry, keeping in mind the general notion of symmetry that we now have in hand. As we'll see in Chapters 7 and 8, its application to more abstract, mathematically defined symmetries had two crucial results. First, it allowed the theorem to become the foundation for our current theories of matter, and, second, it permitted the theorem's application beyond the realm of the physical sciences.

The symmetries that I'm alluding to are more abstract. They do not involve space and time but are related to other properties of objects or systems. To get a taste of what some of these types of symmetries can look like, we'll consider a familiar example involving electricity.

Have you ever noticed a bird perched on a high-voltage power line and wondered how it managed to avoid electrocution? The explanation is that only a *difference* in voltages produces a physical effect. The effect is an electrical *current*, which is the name we give to the movement of charge. All the familiar uses and abuses of electricity are the results of currents flowing through various objects: current flowing through a wire heats it up, the phenomenon that we use to build space heaters, water heaters, and stoves; a changing current passing through a wire creates a magnetic field, a phenomenon that we use to create motors and loudspeakers; sending a current through a thin wire heats it to the point that it glows, leading to the old-fashioned incandescent light bulb.

In these examples, the charge that is in motion takes the form of electrons drifting through metal. In the nervous system, the charges are ions (atoms with more or fewer electrons than protons), and the ion currents contract muscles, send signals to the brain, and, in extreme cases, cause people to write books.

Even a modest electrical current passing through an animal's heart can fatally interrupt its rhythm. Yet the bird chirps on. It can do so because all the currents just mentioned are driven by differences in voltage between two points in space. The sign on the utility pole warns of an alarmingly large number of volts, but this is the voltage difference between the wire and the *ground* on which we stand. If we or something like a ladder we're holding touches the wire while our feet remain on the earth, we're fried as this voltage difference pushes an electrical current through our bodies.

The bird has both feet on the wire and is not in contact with the ground. To be sure, there will be a tiny voltage difference between its two feet because of the nonzero resistance of the conductor, and a consequent tiny current runs through its body. But this minuscule flow of electrons doesn't disturb the creature's blissful ignorance of the laws of electricity.

What I've been calling voltage is a form of energy called the *electrical potential* in the field theory of electricity and magnetism. It's similar to the gravitational potential energy that you gain each time you ascend in an elevator. And, as in the case of gravity, this electrical potential can be converted into an equal quantity of *kinetic energy*, or the energy of motion, as happens when the elevator fails and comes crashing down. (I'm describing an ideal, frictionless tragedy; in reality, some of the potential energy would also be converted into heat, sound, and deformation of materials.)

Here is the connection to symmetry and Noether's theorem: the irrelevance of the *value* of the electrical potential at any point is a type of symmetry. We learned to see the inconsequential nature of the zero of time, or the numbering of the floors of a building, as a form of symmetry: a change or shift in an origin left everything the same. Electrical voltage is another example where the zero point, or the absolute value, of a quantity has no effect and is therefore another form of symmetry.

In the case of time, we know that the symmetry leads, through the theorem, to the law of energy conservation. It turns out that the machinery of the theorem shows that the symmetry in the electrical potential is equivalent to *charge conservation*. This is another basic conservation law, like mass conservation. It had been accepted for a long time as simply another rule for how the universe operates. Charge conservation means that in any system and for any collection of charged objects, the *net* charge (the difference between the total amount of positive charge and the total amount of negative charge) will remain constant in time forever. You cannot create or destroy charge.

Until Noether's theorem, these two fundamental aspects to the universe, the symmetry of voltages and the conservation of charge, appeared as two separate principles. The theorem shows that the two facts about nature are in a deep sense the same fact, a hidden and unimagined connection between two principles that had seemed independent of each other.

The equivalence between electrical potential symmetry and the conservation of charge is yet another surprising and profound insight produced by Noether's result. Without this equivalence, charge conservation, like energy conservation, would be something contingent. We observe charge conservation to be true, but it might have been otherwise. The symmetry in the electrical potential energy is not something contingent on observation but is deeply embedded in the structure of the theory. In showing that charge conservation is mathematically equivalent to potential energy symmetry, Noether's theorem demonstrates that this crucial aspect of observed reality—that charge can be neither created nor destroyed—*cannot be otherwise*. That is, if Maxwell's equations, which define the subject of electricity and magnetism, do describe reality, then this reality must include conservation of charge.

As in Newtonian mechanics, Noether's theorem has again shone a piercing light into an old, well-established theory, illuminating it from the inside. The theorem finally reveals its hidden structure, its hidden symmetries, and its supposed contingencies as necessities and grants us insights that eluded its creator.

Up to this point, we were examining space-time symmetries: time translation, isotropy, and spatial translation. The present discussion of electrical potential considers a symmetry of a similar kind but does not involve space or time. The symmetry in the electrical potential is part of a large category of symmetries with a more mathematical flavor. Physicists call this category *gauge symmetries*. They are symmetries just like all the others: instances of transformations that leave everything as it was. But the transformations have now passed beyond our ancient intuitions about the structure of real-life space and time in which we live and move and into the realms of the abstractions of mathematicians and physicists. And yet intuition is not left completely behind, although our inner suspicions concerning voltage changes, and our lack of *complete* surprise at the oblivious resilience of our feathered friends overhead, is more of an acquired rather than an instinctual condition.

As mentioned earlier, just as Noether's theorem shows that time translation symmetry is equivalent to energy conservation, it also shows that spatial translation symmetry is equivalent to momentum conservation.

The establishment of conservation of momentum in our laboratories today, here on earth, shows that the laws of physics are the same throughout the vast expanse of the universe (aside from complications arising from general relativity). They must be so because spatial translation symmetry means that where we decide to place the zero of our rulers cannot affect the predictions we make about the physical behavior of the cosmos or anything in it.

Various scientists and philosophers have, on occasion, suggested that we might consider an alteration of the laws of physics as we travel through space—that a different physics might hold in different parts of the universe. They point out that, after all, our science is based on observations in the corner of the cosmos where we happen to live. Why should we assume that the laws of physics are the same millions of light-years away?

Noether's theorem tells us why. As long as momentum conservation holds, the theorem proves that the laws of physics cannot depend on where we are—spatial translation symmetry demands that "where we are" has no meaning. This is another example of the power of her mathematical result to bring clarity to questions of what is and is not possible, what we might spend time and energy musing about, and which ideas are simply nonstarters. Here, as in many other places, the theorem, a demonstration in pure mathematics, has far-reaching philosophical implications.

In his book *A Beautiful Question: Finding Nature's Deep Design*, renowned physicist Frank Wilczek makes a related and intriguing physico-philosophical observation about what he calls the "uniformity of substance."[35] He points out something that many people, including many other physicists, have probably never thought about critically, because it's something that we take for granted. It is simply that every example of a particular elementary particle, such as an electron, is exactly the same thing. Different electrons may be in different states—for example, moving with different speeds—but they all have the same properties. There's no such thing as an electron that's a little bigger or heavier than another under the same conditions or one that has a slightly different charge. An electron is an electron; there is no way to distinguish one from another. If you were to go to the most distant corner of the universe and do an experiment on an electron, you would get all the same results as you would on earth.

The fundamental reason for this "uniformity of substance" is that the different electrons are not really separate particles, according to modern quantum theory. They are, instead, "excitations" of a quantum field that permeates all of space. This quantum field is part of the physical description of reality; if energy and momentum conservation hold, then Noether's theorem tells us that because of the symmetries of time and space translation, our description of physical reality must not vary in time or location. The same excitations of the quantum field must occur everywhere, and so all electrons must be identical. Through this argument, Wilczek shows how Noether's theorem not only provides a deep explanation for the uniformity of particle properties but also assures us that these properties must be the same in all parts of the universe. This argument constrains the range of admissible cosmological theories. Whatever ideas cosmologists come up with to explain deep space or the history and structure of the universe, they should assume not only that the laws of nature are the same everywhere but also that the "stuff" that these laws act on is also identical in all parts of the cosmos.

Noether's theorem elevates space and time themselves into active participants in reality, rather than the blank canvas on which physical things happen. The structure of space and time, and the symmetries that hold within it, give rise to the conservation laws that describe the behavior of physical reality. Something very like this is true of general relativity, which transformed our understanding of space and time into a kind of substance that gives rise to the phenomenon of gravity. But Noether's theorem goes further, connecting all physical forces— electromagnetism, the strong nuclear force, and everything else—to a set of eternal symmetries.

Einstein and Energy Conservation
After Noether's Theorem

After some recovery time from his race to make the equations of general relativity work, Einstein returned to open questions related to the new theory. It turned out he had not forgotten about the energy conservation puzzles. Before receiving a copy of Noether's 1918 paper in which she proved her theorem and which impressed him greatly, he was still not completely at a loss. Hilbert had been corresponding with him on the energy conservation issue, as he, Klein, and Noether were gradually coming to a more solid understanding. Or probably more likely, as Noether was working it out or already had done so, she was explaining her results to the two men.

In a letter from Einstein to Hilbert in May 1916, the physicist confesses that he's confused about a version of an energy theorem that Hilbert had sent him. Einstein goes on to say, "It would suffice, of course, if you would charge Miss Noether with explaining this to me."[36]

As much correspondence is lost, we extract whatever insight we can from what survives. Several things are clear from Einstein's remark. It seems there was already, probably beginning soon after Einstein returned from his Göttingen lectures to his home in Berlin, established communication between him and Noether. We can surmise, from the tenor of Einstein's comment and his other letters, that he had already become accustomed to Miss Noether in the role of his tutor, especially in regard to variational forms of the field equations, energy theorems, and related generalizations. It should not be forgotten that this period includes those months where many historians describe Einstein as working in solitude in the search for the correct field equations.

Apparently Einstein reached a better understanding of some of these issues by the end of the year. Almost exactly one year after publishing his final version of general relativity, he published a paper titled "Hamilton's Principle and the General Theory of Relativity." In this paper, Einstein derives the gravitational equations using the variational calculus, similarly to how Hilbert derived them in his own publication. "Hamilton's Principle" in the title is a reference to the methods of the variational calculus. He mentions Hilbert's derivation but points out that his own is better, because, unlike Hilbert, he makes no assumptions about the composition of matter. He also, under certain restrictions ("The energy components cannot be split into two separate parts such that one belongs to the gravitational field and the other to matter, unless one makes special assumptions"), derives a conservation law combining energy and momentum.

Einstein's derivation of this conservation law is a very special case of Noether's theorem that would be published later. It's nearly impossible to avoid the conclusion that he was led there in good part by Noether's correspondence and her tutelage, both directly and relayed through the letters of Hilbert and Klein. He mentions Noether nowhere in the paper, but we should remember that nothing on this subject had yet been published formally under her name. Nevertheless, Einstein's parsimony in mentioning the contributions of others when publishing his works on gravitation has contributed greatly to the myth that he developed general relativity on his own. His preference for sole authorship of these papers also did nothing to counteract the erasure of Emmy Noether's contributions to the history of physics.

Einstein seemed to be proud of this paper, sending copies of it to several colleagues. He was not aware of what would be its most enduring legacy. In this article, finally wearying of the typographical prolixity demanded by the tensor calculus, with its profusion of indices, little subscripts, and superscripts referring to coordinates, Einstein invents a clever notational innovation that makes the equations more streamlined. The scheme was universally adopted with relief and is now known as the *Einstein summation convention*. This innovation alone would have ensured the physicist's immortality.

Continuing their direct correspondence, Noether sent Einstein a copy of her Noether's theorem paper shortly before its formal publication. He reacted with another letter to Hilbert on May 24, 1918, writing with wonderment at what happened to his equations in Noether's hands, how he had never imagined that things could be expressed with such elegance and generality. He teased the others, suggesting that they could have learned a lot from her, had they brought her aboard sooner: "Yesterday I received a very interesting paper by Ms. Noether about the generation of invariants. It impresses me that these things can be surveyed from such a general point of view. It would not have harmed the Göttingen old guard to have been sent to Miss Noether for schooling. She seems to know her trade well!"[37]

Apparently it took Einstein some time to absorb the implications of Noether's paper. A note he wrote to Klein on May 28, 1918, shows that he was still struggling to fully understand energy conservation in general relativity, saying that there was still a "mathematical gap" in his argument.[38] He had grappled with the issue in an earlier incarnation of the theory as well.[39]

One final note about the circumstances of Noether's publication of her paper that has been the subject of this chapter: it was not published in one of the journals in which she had published her previous research, but in the *Proceedings of the Royal Göttingen Academy of Science*.[40] The academy was where Göttingen mathematicians and other scholars liked to give talks and have discussions. Much important research had appeared, and would appear, in the pages of its *Proceedings*.

The Royal Academy had not permitted Noether to join or to present papers. Therefore, her paper was presented on her behalf by Felix Klein. (It was discovered recently that in the early 1930s, Hermann Weyl had tried to get Noether elected into the academy but gave up when it was clear the institution would not budge.[41])

One day, Hilbert decided to punish the members of the academy for their obstinate refusal to accept his colleague, employing a withering, ironic insult. Sitting among them in their hallowed halls, he blurted out, "It is time that we begin to elect some people of *real stature* to this society."

Having gotten their attention, he added, "Now, how many people of *stature* have we indeed elected in the past few years?"

Then he made a point of peering slowly around the room, letting his gaze fall deliberately upon a specific handful of the members there. And he answered his own question: "Zero."

4

Emmy Noether Gets a Job

"How easy this decision would be for us if this were a man."

The Long Road to Habilitation

Let us now move back in time a few years from Emmy Noether's triumph of 1918. She had just arrived at Göttingen and dove happily into mathematical research alongside Hilbert, Klein, and colleagues while she continued to travel to Erlangen regularly to handle family matters. As soon as she arrived at Göttingen, Hilbert and Klein, but especially Hilbert, began a campaign to get her hired onto the faculty, for Noether was working without pay or position.

Germany was one of the last holdouts in Europe to allow women to be officially enrolled as students in universities.[1] France had opened the doors in 1861, England in 1878, and Italy in 1885. An (eventually disputed) account in a Berlin newspaper in 1895 about two professors who had demanded that the females in their classes leave led to a survey of faculty all over Germany about their attitudes toward women being admitted to higher education. The responses followed a peculiar pattern that we will encounter repeatedly. The faculties of history, theology, and similar subjects that we now group with the humanities remained steadfastly conservative, believing that women should stay out of the university classrooms (they were not sufficiently equipped mentally, they should stick to raising children and taking care of the household, their presence would disrupt the university environment, and so on). In contrast, while the faculties of science and mathematics were not yet universally ready to throw open the gates, they were generally more receptive to the idea of coeducation. Those gates opened first in the state of Baden in 1901 and throughout the rest of Germany in 1908, the year that Noether was to receive her doctorate.

Although women were now allowed to matriculate at German universities and receive degrees, they were still banned, by policy of the Ministry of Education, from being hired as teachers at the university level.

In Germany, the "right to teach" at a university had to be earned through a process that resembled a second PhD, called the *habilitation*. This requirement emerged in the nineteenth century and remained in force until quite recently. To habilitate, you submitted a thesis or other academic papers to demonstrate the high quality of your original research; sometimes the application was supported by a lecture. If the committee judged your application to be sufficient, you were then elevated to the position of privatdozent. This position gave you the right to teach at German universities in the capacity that the name suggests: as a private teacher. A privatdozent received no salary. He (always *he* in those days) was allowed to receive course fees directly from students who chose to attend his lectures. This arrangement suggests that the German university system was quite different from what people in, for example, the United States or most of the rest of Europe are used to. There were few required courses and no examinations in Germany, for example. Students were largely free to take whatever courses interested them and even to roam among universities; teachers had to compete for their attention.

The arrangement also reminds us of the class distinctions that were built into the system. Only those with private means could afford to be privatdozents, as the course fees were not enough to live on. While in some sense a meritocracy with a degree of academic freedom that would be unfamiliar in, say, the United States (even more so in recent years), the German university was also an elitist club.

The battle for Noether's habilitation echoed Klein's battle to bring Hilbert to Göttingen, with the baton now passed from Klein to Hilbert, although Klein remained in the race throughout. The differences in attitudes between professors in the humanities and the sciences would soon drive an official division: the faculties would formally split into two schools as a result of a series of conflicts, including those around Noether's habilitation. As might have been predicted, the faction that had opposed Hilbert's hiring at Göttingen would be the lead opposition against Noether's habilitation. The scientists and mathematicians would be dubious about allowing a woman to teach but would be more open to making an exception.

And an exception is what it would have to be. To avoid alarming the faculty and the Ministry of Education, Noether's supporters framed their request to allow Noether to habilitate as a request for a rare exception to the rule and most assuredly *not* as a request to change the rule itself. They were not asking that the doors be opened to female teachers. They were arguing that Noether's singular talents and importance to mathematics made it imperative that she be allowed to teach.

The first try at habilitation occurred in July 1915.[2] With Klein's and Hilbert's encouragement, Noether sent her recent paper "Fields and Systems of Rational Functions" to the Mathematical and Natural Sciences Department of the Philosophical Faculty of the University of Göttingen as part of her official application.

With commendable German bureaucratic efficiency, the faculty called a meeting the very day they received the application and formed a commission to write reports. In a letter full of praise for the candidate, Klein wrote to the Ministry of Education personally, recalling how amazed he was at her mathematical knowledge when, in 1914, she had visited Göttingen with her father on the mission to write an obituary for Paul Gordan. Klein made the point in his letter that Emmy Noether happened to be superior to the average of the *other* candidates who had been passed in recent years.

All the reports, except for one, written by various commission members were in favor of Noether's acceptance *in spite of* her being a woman. They agreed, implicitly or explicitly, that women in general were not suited to be university teachers but that Noether was a singular, special case.

The exception was Hilbert, who had been named "principal reporter." He alone omitted any mention of her sex, writing only about her scientific qualifications.

Under normal circumstances, the reports would have been forwarded up the ladder and to the ministry, getting an automatic rubber stamp at each step. But the sex of the candidate made the circumstances other than normal. One of the commission members, mathematician Edmund Landau, after acknowledging Noether's scientific talent and promise and her skills as a lecturer, lamented, "How easy this decision would be for us if this were a man. . . . I would much prefer it, if this extension of our teaching program could be accomplished without the habilitation of a lady." Landau, who was the department chairman, also shared his view that "the female brain is unsuitable for mathematical production, but I regard Miss N as one of the rare exceptions."

The habilitation commission met again at the end of October to take a vote. With one exception (astronomer Johannes Hartmann), everyone voted in favor of Noether's acceptance.[3]

Because of the unprecedented nature of the case, the entire faculty, including both factions, met on October 18 to have what started as an uncomfortable discussion and got worse from there. Records of the meeting are sketchy, but accounts of Hilbert's explosive derision have survived. This is the meeting at which he turned on the humanities faculty and uttered his legendary remonstrance that the sex of the candidate should not be an issue, because, "gentlemen, we're in a university and not a bathing establishment."

(Hilbert became so notorious for his campaigns for the rights of female scholars that, as a prank on his fiftieth birthday, his friends named him president of the nonexistent "Union of Women Students."[4])

Hilbert wound up having to write letters of apology to two philologists whom he had personally attacked, after they had complained in writing to the department chairman about his "uncollegial behavior."[5]

Apparently, during the meeting, Hilbert had suggested that these two professors should leave. In his coerced apology, Hilbert claimed that he hadn't intended to expel anyone but was merely trying to make the point that politics should remain outside the university's walls and that the faculty should concern itself exclusively with the candidate's merit as a scholar and teacher.

Despite these fireworks, the faculty voted, narrowly, not to stand in the way of forwarding the habilitation application to the ministry. Hilbert, however, was realistic about its outcome there. He had already arranged, through a back-channel communication, to have a sit-down with the ministry. So he was in the room with the ministry members soon after they decided to pocket veto Noether's application, setting it aside without officially rejecting it. He made a deal with ministry officials to allow Noether to teach classes that would be officially listed under Hilbert's name—of course, with no pay. She started teaching classes almost immediately as a Hilbert stand-in, surviving by living frugally on a small inheritance.

The second try took place over a year later, during the summer of 1917. The Göttingen mathematicians had heard a rumor that Frankfurt

University would try to poach their Emmy Noether and somehow finagle her habilitation from there.

The reply was surprisingly rapid. The minister reassured them that they need not worry: Noether would not be allowed to habilitate at Frankfurt, so they were in no danger of losing her because of that. Nor would she be allowed to habilitate at Göttingen or anywhere else, in case they were still wondering.

The stage was set for the third try by the end of World War I and the collapse of the German monarchy. There was a sudden liberalization of society, and opportunities began opening up for women. It was at this point that Albert Einstein became personally interested in Noether's case.

Although Einstein was not yet the international superstar that he would shortly become, he already had considerable juice within the academic world, and he appeared to know it. He had received a copy of the published version of the Noether's theorem paper, which gave him another opportunity to try to absorb its results and implications. It made an impression on him. Shortly after the end of the war, he wrote to Felix Klein: "After receiving the new paper by Frl. Noether I again felt it is a great injustice that she is denied the [habilitation]. I would very much favor that we take energetic steps with the Ministry. If you do not regard this as possible, then I will make the efforts myself."[6]

Klein had no disagreement. In fact, he said that her output as an unpaid, unofficial assistant, just in the previous year, had been superior to anyone else's, including the full professors in the department—including himself and Hilbert.

The university sent another application to the ministry, including the Noether's theorem paper and another paper as supporting documents.[7] Noether's summary in the application, reflecting her customary modesty and understatement regarding her own work, said that both papers were "in part, the result of the assistance that I provided to Klein and Hilbert in their work on Einstein's general theory of relativity."

Scholars concur that the "assistance" Noether refers to here resulted from Hilbert's and Klein's consultation with her on their puzzlement over the apparent lack of energy conservation in Einstein's theory.

Noether's supporters were hopeful because the ministry leadership was now in the hands of people with far different political leanings than before. German women now even had the right to vote (a bit before women would win that right in the United States).

Nevertheless, the 1908 policy prohibiting women from entering German university faculties had never officially been repealed. But on June 4, 1919, the Ministry of Education approved Noether's habilitation as an exception.

One year later, the ministry finally repealed its no-women policy, in response to a petition from Göttingen phenomenologist Edith Stein, who was murdered by the Nazis in 1942 and canonized by the Catholic Church in 1998 (the former because of her Jewish origins, and the latter made possible by her conversion to Catholicism).[8] Stein was a cousin of mathematician Richard Courant, who played a role in the promulgation of Noether's results, as we'll see in later chapters. On February 21, 1920, the ministry decreed that "belonging to the female sex may not be regarded as an impediment to receiving the Habilitation."[9]

It's important, in trying to understand the past, to avoid anachronistic thinking. If we sincerely want to grasp the motivations and actions of people of another era, we must not transport our attitudes and assumptions into a time where they do not apply. Of course, we cannot and should not avoid holding figures from the past to account by the lights of our own moral standards. Monstrous acts of history remain monstrous, no matter how well accepted they may have been at the time. Without the willingness to pass such judgments, we could not learn from history or measure the progress of justice and equality. My point is more immediate: unless we understand the worldviews of the inhabitants of earlier eras, no matter what we think of them, what they did and said won't make sense. We won't *get it.*

To this end, notice what Hilbert *did not say.* He did not say that Noether should be habilitated because of human rights, or women's rights, or equality, or fairness. He did not say that sexual discrimination was wrong or immoral. He did not use the language of rights or equality.

Hilbert argued that Noether's sex was *irrelevant*: this is not a bathhouse. It was good for science to hire the best people, and she was the best person. Nothing else should matter.

Hilbert was completely alone in his stance. Every other defender of Noether's habilitation advocated it *in spite of* the sex of the candidate. As we've seen, many were at pains to emphasize that Noether was a special case that should be treated as an exception, and her habilitation was definitely not to be a change of policy.

For Hilbert, any policy that stood in the way of the progress of science was wrong because of that, not because it was unfair to individuals.

(In this connection it is worth noting Einstein's interesting use of the term *injustice* in his letter supporting Noether's habilitation. He strikes a chord here that her other supporters never played. Even more than Hilbert, Einstein was a man out of his time. His keen sense of social justice would find a passionate outlet for expression after his escape from Nazism and his move to the United States, where he encountered another form of fascism. Einstein's steadfast activism on behalf of the rights of African Americans and his condemnation of segregation reveals a great deal about his unique character.[10])

Hilbert was never particularly scrupulous about giving Noether credit for her work or, for that matter, overly concerned with amassing credit for himself. Remember his casual readiness to disavow any responsibility for the creation of general relativity. His single-minded devotion to truth and to the progress of science and mathematics—principles for which he was ready, without a second thought, to place his career and personal reputation in jeopardy—was his only motivation. I see no convincing evidence that Hilbert was very much concerned with helping Noether as a person, either before or after her habilitation. The evidence is overwhelming that he saw her as someone with a rare ability to push science forward, that nothing else mattered, and that he simply could not understand why anyone was talking about an irrelevant characteristic.

Hilbert embodied the Platonic ideal of the meritocracy. Not because it was right, but because it *worked*.

Emmy Noether was nearly the only, but not *the* only woman mathematician in Germany during her time.

The story of Grace Chisholm Young is especially interesting.[11] After a spectacular performance in mathematics at both Oxford and Cambridge (even though these universities did not yet confer degrees upon women), the English mathematician decided to enter the big leagues and enrolled at Göttingen in 1893. Although German universities were not yet admitting women in general as official students, the Ministry of Education decided to make an exception that year and allowed Young to matriculate, along with two American women. These three were the first women officially enrolled in a German university as degree candidates. Young received a doctorate (magna cum laude) from Göttingen under Felix Klein in 1895: the first PhD awarded to any woman in Germany (not counting the doctorate in mathematics from Göttingen awarded in absentia to Sofya Vasilyevna Kovalevskaya in 1874). Young then returned to England. None of her publications during this period are in German journals.

Margarete Kahn and Klara Löbenstein, who were friends, began auditing mathematics at Berlin and Göttingen around 1905, obtaining their doctorates in 1909 under Hilbert, with the objections of the Berlin faculty defeated by Hilbert and Klein. (Kahn's and Löbenstein's thesis defenses were on the same day, June 30, 1909.) As there was no way yet for women to pursue academic mathematics as a paying job, both women got positions as schoolteachers. Aside from their dissertations, preserved in the Göttingen archives, they had no publications.

Noether was the only female mathematician who managed to habilitate in Germany during her time. Hilda Pollaczek-Geiringer earned her habilitation in 1927, but she fled the Nazis before she could take up her job.

As Noether showed, you didn't have to be a mathematics professor to publish papers, but having to either be a housewife or make a living some other way would leave little time for the rigors of research in pure mathematics that would lead to publication. Noether had the freedom to pursue research because of her small inheritance, extreme frugality, and lack of interest in marriage or much else outside of mathematics.

Although the obstacles in the way of a woman interested in pursuing mathematics as a career were, obviously, formidable at the time, they loomed even higher in the areas of research more closely associated with test tubes and lab coats. As Claire G. Jones points out, mathematics did not "professionalize" in the same way as the laboratory sciences, in whose physical arenas, marked by ranks of more formally delineated apprentices, technicians, and senior researchers, all bumping shoulders in environments that were often intense and could be hazardous, "any hint of femininity came to be seen as incongruous and inappropriate."[12] Jones also makes the interesting observation that around 1900 there were proportionally more women working as research mathematicians than in the 1960s.[13]

Noether's Boys

Having dispensed with Hilbert's and Einstein's physics issues, Noether now turned her attention to pure mathematics. The ongoing debates about the meaning and interpretation of energy and anything else in general relativity continued without her. For the rest of her life, she only referred to the Noether's theorem paper once, in a brief note. It had served her well, helping to push along her habilitation, but she had no particular interest in the physics problems that had motivated it.

She taught for a couple of years as a privatdozent. Then, on April 6, 1922, the Minister of Science, Art and Public Education (the new title of the position) promoted her to the lowest professorial rank.[14] This was something of a concession for the previous delays in approving her habilitation, as it customarily took much longer to ascend from the lowly position of the privatdozent. In a way, they were giving Noether credit for "time served."[15]

However, attitudes had not changed completely from earlier times. Her promotion came with the stipulations that her new position would not come with a salary, that she would not be a member of the civil service and therefore not receive any benefits or pension, and that (quite remarkably) her administrative position relative to the other faculty members would be *unchanged* from that of her previous position as a privatdozent.[16] This last stipulation seemed to be intended to make it clear and official that this woman would not be granted any form of authority over any male employee.

The bottom line was that she had been given a position with a title but little else. However, there was one thing: with the job came the right to supervise PhD students.

There followed a curious development. Despite the warning in the letter from the ministry, Hilbert requested and received a contract from the university to pay Noether to teach algebra classes. It seems that he had neglected to inform the university administration that she wasn't supposed to be paid by the government.[17]

The salary was minuscule, not quite enough to live on, but she was happy to receive it. Her friends even convinced her to celebrate her sudden wealth by taking the radical step of buying some new clothes.

In the small worldwide community of mathematicians, Noether rapidly became well known as someone making increasingly original contributions to a variety of subjects, although many who read the papers of an "E. Noether" probably knew nothing of the author as a person. All those who did know her and who left behind a record of their experiences tell a similar story. Emmy Noether could make an unsettling first impression. She spoke loudly and laughed even more loudly. Her manner was abrupt and appeared, to her compatriots, as unfeminine. She didn't hesitate to interrupt you to tell you that you were wrong. When attending a lecture, she was capable of jumping out of her seat to grab the chalk and take over if she thought of a better way of doing things. But those who survived the initial encounter universally admired her and grew to like her. It quickly became clear that she intended no rudeness; it was just that, to Noether, math was more important than manners.

She adopted with pleasure the Göttingen tradition of hashing out mathematics while walking around the town and the forest. One of her colleagues put this predilection to his own use, suggesting a nice walk when she wanted to talk math. This was the only way he could slow down her rapid-fire delivery.[18]

As Hilbert had done before her, Noether now also began to accumulate a collection of widely repeated anecdotes and gained a reputation for her own brand of eccentricity. Although she became a teacher beloved by her advanced students, who were devoted to her, she could be a disaster in the classroom for beginners or, in general, for those without the fortitude to follow a complex train of thought, as a mathematical genius invented new ways to do things, right on the blackboard in front of a crowd of students—some transfixed, and some overwhelmed to the point of running out of the room.

"Running out of the room" is not an exaggeration. It was such a regular occurrence that the band of talented young mathematicians known as Noether's Boys, who had devotedly attached themselves to her, made a game out of it. When they spied any new student in her classroom, they would wait with gleeful anticipation, watching as he shrank in terror under the force of the breakneck intensity of her eccentric lecturing style. When he escaped, usually after about ten minutes of the onslaught, they would cheer that victory was theirs, as the enemy had again retreated.

This aspect of Noether's lecturing style shared something essential with Hilbert's, who was also fond of working things out on the blackboard, on the spot. But Noether brought her own brand of frenetic passion to her teaching. She did not rehash material from textbooks on the blackboard. She improvised, finding new ways to explain things and inventing new proofs for theorems as she stood in front of her students. She did all this at a rapid-fire pace that required the utmost concentration to follow but that led those who could manage on a journey of discovery that they found delightful. She was showing them how to be mathematicians.

Only the most talented and confident mathematicians can successfully adopt such a seat-of-the-pants approach, but even among this elite group, it's a minority practice. Klein personified the opposite style: his lectures were famous for their impressive degree of organization. At the end of one of his classes, the blackboard was as carefully and clearly laid out as a textbook.

Sometimes, Noether's passion got the better of her, as when her exploratory method led to a dead end. On at least one such occasion, she threw her chalk to the floor and stomped on it, cursing her poor treatment at the hands of the mathematical gods.[19] Even when things were going well, she was so caught up in the whitewater rapids of her lecturing that by the end of the session, her hair had come undone and was flying about her head, and her clothing was in disarray. When things got so bad that the few female students in attendance could take it no more, they would run to the front of the classroom to help their professor put herself back together, but after a few seconds of this, she would shoo them away.

Not everyone partook in the joyful and heady atmosphere around Noether at Göttingen in the 1920s. It probably won't be surprising to learn that a certain strain of jealousy erupted among a gang of young lecturers who envied her fame and her capture of the best students in the department.[20] This envy was no doubt intensified by their reaction to the unprecedented phenomenon of a woman becoming the most important mathematician in town. It found expression in sarcastic jibes directed at the professor and at Noether's Boys.

Despite her fearsome reputation as a teacher, Noether was capable of giving a user-friendly lecture when the occasion called for it. Eventually, she become so well known among mathematicians and students that she might enter the lecture hall to see a hundred people, including visitors from Holland, Russia, and elsewhere, assembled to hear her. After checking to see that they hadn't entered her room by mistake, she would give a lecture of organized clarity that all could understand.

There was a long tradition in German scholarship of "schools" arising around an influential philosopher, scientist, or mathematician. The school consisted of a group of students and fellow thinkers through whom the leader promulgated his ideas, results, and style of analysis.

During the 1920s, a school of mathematics grew up around Noether, who became, almost certainly, the first woman in German history to be the leader of such a group. Here she followed in the tradition of Klein and Hilbert, who themselves were leaders of highly important schools of mathematicians. Through her school, Noether became one of the most important mathematicians of her time. She generated a series of her own timeless results in the field that she was remaking: abstract algebra. She taught others her methods of working, which she claimed were ultimately more important than her actual results. She launched many careers, sometimes giving a start to a young researcher by making a gift of a result that she had not had time to publish or simply hadn't bothered to do so. On such occasions, she would insist that her associate develop the idea further and publish it, without mentioning or crediting her.

This last form of generosity will probably seem astonishing to us, especially if we are familiar with academic life and the jealous guarding of credit for every trivial bit of work. But Noether's imagination was so bursting with fertile ideas that she may not have had the chance to develop them all into papers. Her interest in her own standing, while not entirely absent, was subservient to her impulse to help her students and younger colleagues, and more than one observer remarked on her "motherly" concern for the young people under her wing.

One anecdote in particular provides a perfect example of this impulse.[21] One day, one of Noether's young associates, B. L. van der Waerden, got a result that he was excited about. He wrote it up in a paper that he showed to her. She told him that it was good work and that she wanted to send it off for publication. But first, she showed him how he could improve the organization of the paper. He did so, and the paper appeared under his own name, with a review by Noether herself.[22]

What van der Waerden learned later from another student of Noether's was that not only had she reached the same result long before he had, but she had used it in lectures that this student had attended at Göttingen. Noether had never bothered to publish it and had never said a word about it to van der Waerden. When van der Waerden told this story, which he loved to repeat, in later years, he would say, "She did not want to spoil the young man's joy over his discovery! Isn't that fabulous?" (In Chapter 6, we'll learn how van der Waerden became the most important disseminator of Noether's results in algebra.)

Noether's only female PhD student at Göttingen turned out to be an important scholar with a fascinating career. In some ways, Grete Hermann's early life echoed Noether's. She too was born and raised in Germany and became qualified to be a schoolteacher but suddenly decided to study mathematics instead. And she too was supported in her ambitions by her father, whose motto was "I train my children in freedom!"[23]

Hermann matriculated at Göttingen as a student of mathematics, physics, and philosophy in 1921. She would go on to do important work in all three fields, finding unique synergies among them.

After Hermann received her doctorate for work that would become important later in the field of computer algebra, Noether was making efforts to find her a mathematics job.[24] But when Hermann decided to pursue a career in philosophy instead, an exasperated Noether was prompted to complain, "She studies mathematics for four years, and suddenly she discovers her philosophical heart!"[25]

Hermann would make contributions to several areas of philosophy, but she is chiefly remembered today for her work in the philosophy of quantum mechanics. Her ideas and papers in this area were studied by some of the founders of this branch of physics, including Heisenberg.

An interesting episode from her career concerns the overturning of a result published by John von Neumann and having to do with quantum mechanics. Von Neumann was a Hungarian-born mathematician who had worked with Hilbert and eventually settled in the United States in 1930. He was an astonishing child prodigy who made important contributions to nearly every area of mathematics. His intense interest in applied mathematics led to fundamental contributions in physics and to the establishment of much of the theoretical basis for digital computers. Such was von Neumann's intimidating reputation that few had the temerity to question his judgment or doubt his results. Hermann has the rare distinction of having proven von Neumann wrong.

Most people are aware that quantum mechanics, the theory of physics that governs the microscopic world and forms the basis of our theory of matter, involves chance in some essential way. But many people are unaware of how deep the role of chance, or probability, goes. In quantum mechanics, at least in accord with its dominant interpretation, the results of measurements of basic events at the microscopic level are governed by probabilities.[26] We cannot, for example, say that a particular atom that forms part of a radioactive element will decay within some specified time span. We can only say that it will do so with a certain probability. The theory of quantum mechanics provides methods to calculate these probabilities.

This state of affairs is disturbing to many, because it amounts to a profound alteration of our ideas of cause and effect. Everything that happens was supposed to happen out of necessity, as an effect following a chain of causes. When probability entered physics, as it certainly had many times before, it appeared as the expression of our uncertain knowledge of these causes. But in quantum mechanics, probability appears at the base level of reality. Probability is not a summary of our incomplete knowledge; it is a replacement for deterministic causation.

Einstein himself, the scientist who essentially founded quantum mechanics, found the role of chance in the theory simply unacceptable. He expressed his disquiet in his famous pronouncement that God did not play dice with the universe.

Einstein was not alone in his misgivings. A series of theoreticians tried to escape the grip of probability in quantum mechanics by proposing the existence of a deeper causal layer lying underneath the randomness. The knobs and switches in this deeper layer were referred to as *hidden variables*: they were hidden because we hadn't found them yet, but they were there, controlling the results of measurements deterministically, just as in classical physics. Once again, probability would be demoted to a description of our incomplete knowledge of these hidden variables.

In 1932, von Neumann had published what he thought was a proof of the *impossibility* of hidden variables in quantum mechanics. He showed that the mathematical structure of quantum theory was inconsistent with the existence of a causal mechanism underlying the randomness—or he thought he had.

Three years later, Hermann noticed that von Neumann's proof contained a mathematical mistake. It was invalid. She published this finding, but it was ignored. For the next thirty years, whenever anyone mentioned the possibility of hidden variables, someone else remembered that von Neumann had proved they were impossible. That shut the idea down, because nobody argued with von Neumann.

That's where the matter stood until 1966, when Irish physicist John Stuart Bell published a paper showing that von Neumann's proof was invalid. This time, the scientific community paid attention, for some reason. Now everybody knew that the great von Neumann had made at least one mistake: a curious episode in the history of science.

As an interesting epilogue (and another example of how Hermann's work was overlooked), the current *Encyclopedia Britannica* entry for von Neumann mentions his flawed proof and its refutation by Bell but does not mention Hermann.[27] And this is the version believed by nearly all who have scanned the history of quantum mechanics: that von Neumann's demonstration was believed to be valid by everyone for three decades, until the Irishman vanquished it. The fact that it was demolished a scant three years after its publication by Grete Hermann is still submerged in obscurity.

In the following chapters, we'll discover how many of the events and circumstances of Noether's life resembled Hermann's. As in the case of so many women scientists, she was not rewarded or recognized during her lifetime to any degree commensurate with the importance of her contributions. The disregard continued after her death, as the history of science was and is written as if, like Hermann, she and her work never existed. This neglect of the contributions of women scientists is not a feature confined to previous generations, but it persists, despite improvements, to the present day.[28] In the case of Noether, the familiar pattern takes on an even deeper significance, for her work was not merely important; it lies at the very foundation of the most fundamental of the sciences. Her erasure from the history of science, therefore, creates a profound distortion in the general understanding of our common intellectual heritage and its development.

5

The Purge

"Mathematics knows no races or geographic boundaries."

The German academic game of competition for professors has figured several times in this story. The personnel files of Göttingen University contain a typical document relating to the constant threat of the poaching of prominent scholars. It's a letter from 1926 by mathematician Richard Courant warning that other institutions may be planning to lure away famous physicist Max Born.[1] Courant urges that Göttingen should make an active effort to retain Born, because he is "one of those colleagues upon whom the international reputation of this University rests."

Near the word "colleagues" an anonymous official has penciled in the word "Jews."

―――――

Up until 1933, the overt discrimination that Emmy Noether faced was based solely on the fact that she was a woman. Her family, although not strongly interested in religion or in their Jewish heritage, was not concerned with hiding it. In fact, as a girl, Emmy had attended Jewish religious school for a while, which was a bit unusual for the time and place. As an adult, she showed no interest in these things.

Both Max Noether and Paul Gordan (Emmy Noether's PhD adviser) were also Jewish. These men were denied opportunities and suffered long delays in receiving appointments, because of their ethnicity.[2] In a foreshadowing of the support that Felix Klein would lend to Max's daughter decades later, the eminence of mathematics helped the father secure his first position at Erlangen in 1875.[3]

Emmy Noether's heritage was not a secret and was easily discovered by anyone with sufficient idle curiosity. Some evidence suggests that her Jewish background might have been the reason for some of the treatment she received, although she would have been unaware of the details. For example, several writers believe that the refusal on the part of the Royal Göttingen Academy of Science to admit Noether as a member was due to three strikes against her: she was Jewish, she was female, and she had been at least friendly with socialists. (Many did consider her a socialist, although she had long since let a brief party membership lapse and did not take part in political activities.) It might very well have been that if any one of these demerits had not existed, then the academy could have been convinced to let her join its ranks.

She was certainly aware of being forced to vacate her apartment in 1932 after being denounced by another resident of her rooming house as a "Marxist Jewess."[4] The eviction is the first such incident involving Emmy Noether that history records. (The *town* of Göttingen reflected a political culture very different from the university, voting for Nazi-affiliated parties in proportions significantly higher than Germany as a whole.)[5]

Göttingen University, at least the parts swarming with mathematicians and scientists, may have been a partial refuge from the antisemitism endemic throughout continental Europe. But that the disease of Jew hatred was endemic, had been for centuries, and could make the lives of Jews trying to pursue a scientific career miserable is beyond debate. In his autobiography, *Quest*, Leopold Infeld, Einstein's collaborator and coauthor of the popular book *The Evolution of Physics*, gives a moving account of the struggle to pursue a scientific career as a Jew in Europe before World War II.[6] Infeld relates the impossibility of attaining an academic post in his native Poland and his repeated rejections, while non-Jews with inferior qualifications enjoyed enviable careers. He describes his near amazement, on traveling to England, to discover a land where he was judged on his merit and where his ethnicity was essentially irrelevant.

Immediately after the end of World War I, a new strain of antisemitism took root in Germany, one that would directly harm Noether and many of her colleagues. In his commentary to the published volume of his correspondence with Einstein, Max Born identified when "the great danger of antisemitism to German science first appeared."[7] He described a meeting of German doctors and scientists in September 1920, where Einstein responded with uncharacteristic anger after being viciously attacked by physicists Philipp Lenard and Johannes Stark. Born traces the absurd invention of the idea of "German" versus "Jewish" physics to Lenard, who began a systematic persecution of Einstein after this meeting. The two antisemitic physicists became scientific administrators under the Nazi regime and were responsible for the purging of Jewish researchers. Although both were important scientists who eventually were awarded the Nobel Prize, Stark and Lenard are as likely to be mentioned today for their collaboration with the Nazis and their persecution of Einstein, including their unhinged attacks on "Jewish physics."

(Five years before this meeting, David Hilbert had blocked Stark's appointment to Göttingen, expending considerable effort and political capital to do so. Hilbert acknowledged that Stark was the most qualified candidate academically but insisted that his nationalism and antisemitism made him unacceptable for Göttingen. I suppose this episode delineates the limits of Hilbert's purely meritocratic policy.)

University students were in some ways the most enthusiastic vanguard of the Nazi pathology, as is so often the case with reactionary movements.[8] By the 1930s, the idea that there were such things as Aryan and Jewish mathematics and physics was widely accepted, with the Jewish kind supposed to be somehow decadent or non-German. An anecdote dating from this period from Göttingen may illustrate the social disease that had swept over the university and set the stage for what was to come. Students organized a boycott of Jewish Göttingen professor Edmund Landau's classes. (We met Professor Landau in the previous chapter, lamenting the inconvenience that Emmy Noether happened to be female.) When he arrived to teach, he found the doors of his classroom blocked. The student organizers explained to him that they were Aryans and insisted on being taught only Aryan mathematics.

Some German academic officials who were responsible for enacting the government's antisemitic policies and who bent their decisions to accord with the popular anti-Jewish hysteria didn't consider themselves antisemitic. Occasionally, Nazi students were also given to rationalizing their actions against their Jewish teachers. One such leading student activist was at pains to explain to Landau the reasons for the boycott against him. He told the professor that the protest was nothing against Landau "as a Jew" but was meant only to "protect" the young German students from having to learn calculus from someone of an "entirely foreign race."[9] He had no doubt that Landau could teach the "purely international scientific aspects" of math, but in this student leader's view, math contained "broader educational" values, something to do with the "spirit" of the thing, which somehow was connected to the "racial composition of the individual." Therefore, clearly, "a German student should not be trained by a Jewish teacher."

Two days after hearing all this, Landau applied for early retirement.

In 1935, an acquaintance of the by-then expatriated mathematician Richard Courant wrote to him, describing how antisemitism had become endemic in Germany and adding, "One has the feeling that the greater part of the population is mad—one of the characteristic features

of such madness being that victims are normal and (frequently) delightful people—on all subjects but two or three."[10]

The idea of racial differences in scientific or mathematical thought
seems ludicrous to us, but this belief was the received wisdom in the
Germany of this era.[11] As I've argued earlier, if we want to understand
the actions and utterances of people from another time and place, we
must make an effort to know their worldviews, no matter how alien they
may seem.

Even the brilliant Felix Klein took the idea of genetically based
differences in mathematical style and ability for granted, freely par-
taking in the amateur anthropology indulged in by his colleagues. In
Klein's case, when speaking of the influx of Jewish mathematicians into
German mathematical faculties, he welcomed the "fresh blood" as an
invigorating tonic, healthy for German math. Of course, this new blood
included Jews who had converted to Christianity. Jewishness, and whatever influence it had on mathematical style, was in the blood and was
unaffected by baptism.

On an individual level, Klein sought appointments for many Jewish
colleagues, sometimes, with embarrassment, discussing with them the
regrettable circumstance of the resistance they would face because of
their race. An institution or a department would often willingly accept
a certain number of Jews but rankle if the number became too large,
obliging Klein to perform delicate feats of demographic balance as he
built his mathematical empire. (I would feel slightly remiss here in not
reminding the reader that many institutions of higher learning in other
countries, including the United States, also enforced similar unofficial
or secret quotas at the time.)[12]

On this question, as on so many others, Hilbert again stands alone.
We may wonder why his words, reproduced below, are so often quoted.
Isn't he stating the obvious? But that's rather like criticizing Shakespeare for using too many clichés. Hilbert spoke these words in a different context, a speech encouraging renewed international cooperation

in mathematics after the end of World War I. But his statement would apply even more poignantly to the situation a few years later:

> Let us consider that we as mathematicians stand on the highest pinnacle of the cultivation of the exact sciences. We have no other choice than to assume this highest place, because all limits, especially national ones, are contrary to the nature of mathematics. It is a complete misunderstanding of our science to construct differences according to peoples and races, and the reasons for which this has been done are very shabby ones. . . . Mathematics knows no races or geographic boundaries. For mathematics, the whole cultural world is a single country.[13]

The German race theories that Klein and his friends took seriously may seem weird and silly. But nothing is more dangerous than an idea. By the 1930s, the academically genteel theories of disparate racial abilities, often applied in praise of the Jewish mathematical mind, or at least neutrally, had rapidly evolved into a vicious racism that saw Jews as invaders and corrupters of good German society, with their decadent math and science to be stamped out in favor of the proper Aryan variety. Actual academic papers appeared in scholarly journals, explicating the corrupting influence of Jewish mathematics and science. In one of these absurd analyses, Hilbert's math was held up as a typical example of the "Germanic type" while Noether's was an example of the other kind.[14] Einstein, now having achieved celebrity status, became a popular whipping boy in the German press. Antisemitic writers leveraged the counterintuitive nature of relativity to mock Einstein as the representative of a corrupt, Jewish science that obviously made no sense. One German magazine carried his photograph on the cover over the words "Not yet hanged."[15]

The ancient prejudice had affected Einstein's prospects in the past in more subtle ways, of which he was sometimes unaware. Einstein achieved his first academic appointment, at the University of Zürich, despite faculty misgivings about his "Israelite" background.[16] They wrote in their report that they might have been concerned about disagreeable Israelite character traits had not the recommendation of one of Einstein's acquaintances reassured them that he was an exception to the common type.

The Nazis were able to weaponize the set of preexisting odd racial notions, combined with the ancient hatreds, to make such measures as the purging of Jews from faculties and from the entire German civil service acceptable to a critical mass of not only ignorant villagers but also the intelligentsia. And the world will never forget, and must never forget, the catalog of horrors that followed.

———

There seemed to be particular resentment in some quarters against the Jewish intellectual class. Jews made up about 1 percent of the population of Germany but about 8 percent of the teachers in German universities. In the days preceding Hitler's takeover of the government, the warring political parties formed paramilitary groups to do battle in the streets, to disrupt each other's rallies, and to protect their own. The Nazi paramilitaries were the Sturmabteilung, also called the SA. They went beyond interparty conflict, enforcing the National Socialist program through violence and intimidation against Jews and their property. Today, their nickname, the Brownshirts, survives as a synonym for fascist violence.

Because Hilbert had retired from Göttingen University in 1930, he was spared having to deal directly with what was to come. He stayed in the town until his death in 1943.

In 1933, when the Nazis gained control over Germany, Emmy Noether was one of the first to be summarily dismissed in what was to become a series of purges that removed all Jews from government and academic positions. Oddly, the order to remove her referred to the new law mandating the separation of Jews from the civil service, but as described earlier,

Noether's appointment specifically stipulated that she would not be one of its members. However, her lowly status failed to protect her.

Some academics, seeing what was coming, had already left Germany. Einstein was already gone when the Nazis took over. They stole his bank account and properties, claiming to have discovered a dangerous weapon in his vacation house (it turned out to be a bread knife).[17]

The distaste for Germany would remain with Einstein for the rest of his life. He even condemned his friend Max Born's return to Germany in 1953 "to help with the country's democratic rehabilitation" and criticized others who decided to return after the war.[18]

Noether's friend the mathematician Pavel Alexandrov would later describe the event this way:

> German culture and, in particular, Göttingen University, which had nurtured that culture for centuries, were struck by the catastrophe of the fascist takeover, which in a matter of weeks scattered to the winds everything that had been painstakingly created over the years. What occurred was one of the greatest tragedies that had befallen human culture since the time of the Renaissance, a tragedy which only a few years ago had seemed impossible in twentieth-century Europe. One of its many victims was Emmy Noether's Göttingen school of algebraists. The leader of the school was driven from the halls of the university.[19]

After her expulsion, the mathematician Helmut Hasse organized a letter-writing campaign supporting her.[20] He persuaded fourteen colleagues to write about Noether's importance and argued that she should be allowed to retain at least a small group of advanced students. The government disagreed.

Hasse presents an interesting, if disturbing, case study of the psychological conflicts festering in the minds of some German intellectuals at the time. This support of Noether, and Hasse's other relations with Jews throughout his career, make it clear that he didn't partake in the slightest in the antisemitism of those in power. However, Hasse's nationalistic fervor caused him to support Hitler as a hero and to aspire to join the Nazi Party. Nazi ideology was undoubtedly rendered more palatable for Hasse because of some of his peculiar ideas. For example, he once declared that American slavery was a good thing for black people, a judgment that he treated his acquaintances to when visiting Ohio State University in the 1960s.[21]

Other members of the academic community at Göttingen were less subtle in their support for Nazi policies. Werner Weber, whose doctoral dissertation was supervised by Noether, praised her expulsion by the Nazis and enthusiastically cheered the expulsion of others, including Landau, who was also on his thesis committee.[22]

Even after her purging, Noether kept her spirits up to the extent that she inspired those around her. Hermann Weyl remembers that "her courage, her frankness, her unconcern about her own fate, her conciliatory spirit, were in the midst of all the hatred and meanness, despair and sorrow surrounding us, a moral solace."[23]

Weyl had taken over Hilbert's position as head of the Göttingen mathematicians. By now, Weyl, Noether's contemporary, had become a famous mathematician, theoretical physicist, and philosopher. When he was offered the prestigious appointment at Göttingen, he initially refused, saying that he would be embarrassed to take a position that should rightfully belong to Noether, whom he regarded as far superior to him in mathematical talent. But Noether told him that he was being silly: the university would never offer this position to a woman, and if he didn't take it, someone less worthy would. So, with her blessing, he stepped into the job. Weyl would later write of the Nazis that he was "deeply revolted by the shame which this regime had brought to the German name."

Immediately after the events of 1933, he quit his job. Although he was not Jewish, his wife was, and this may have hastened his decision. He had previously received offers to come to America and join the new

Institute for Advanced Study in Princeton, New Jersey, but he couldn't bear to abandon his beloved Germany, despite his horror at the worsening political landscape. Now, he prepared to leave.

There are several theories attempting to explain why Noether was such a priority for the Nazis. Fierce misogyny was deeply rooted in Nazi ideology, beginning with Hitler's remarks about the nature of women in *Mein Kampf.* Noether would have stood out as a rare female member of the faculty of an important university. Göttingen University itself was high on the Nazi list of enemy camps, because of its reputation as not only liberal in outlook but also welcoming of foreigners of all kinds and of women. As mentioned, Noether had, at times in the past, been a member of the Social Democratic Party (a leftist party hated by the Nazis) and had even lent her apartment to some of the members for political meetings. She had expressed pacifist ideas during the war. She had been a visiting professor in Moscow in 1928–1929 and, on her return, had made approving remarks about the people and society there. Some writers have claimed that Noether's early purge had less or even nothing to do with her identity as a Jew but was instead entirely due to the Nazi authorities' perception of her as a Communist or a Communist sympathizer. However, her official notice of dismissal mentions her Jewish background. Nevertheless, her political sympathies and her sex undoubtedly played a big part in putting her near the top of the list.

Being unemployed had never stopped Noether from doing and teaching math before, and she was not about to let the lack of a job slow her down this time, either. But this time was different. There would be no lecturing under the radar in Göttingen classrooms. There would be no appeal. Not even Hilbert could help her now. It was one thing to thumb one's nose at a gaggle of stodgy professors of theology; one could even enjoy watching their eyes widen behind their monocles. But now people were being shot in the street. There would be no defying the Nazis.

And yet, she did defy them. On her removal, she immediately did two things. First, thinking, as always, of others rather than worrying about herself, she joined Weyl (who deferred his escape to Princeton) and immediately organized to help other purged math professors survive. Second, she moved her classroom to her apartment. She had found no legal loophole: she was not permitted to teach Göttingen students at all, with or without pay, officially or unofficially. These math meetings were secret. Noether and her students knew that this was a subversion of authority. But she loved her students and would not leave them, and they loved her, and they flocked to her little apartment for regular seminars, and to sample her sometimes-amusing attempts at cooking.

One day, after most of Noether's students had arrived at her apartment for another secret class, there was a knock at the door. She opened it to reveal a man dressed in the uniform of the brutal Nazi SA: the Brownshirts.

Everything froze, and time stopped.

Then suddenly the world was reanimated with greetings, smiles, and warm embraces. The man entered the apartment, his jackboots thumping the floor, and took his place with her other students.[24]

Noether was permanently ready to joyfully discuss mathematics with anyone, at any time, under any circumstances. Math was her life, her only real interest. For most people, the daily concerns of Noether and those few of her comrades who traveled with her at the highest reaches of mathematical thought are part of another, barely perceivable dimension. She lived her life as if, to her, the normal affairs that occupy people's days—politics, relationships, material comforts—were equally murky and distant and of no real importance to her.

After the purges of the Jewish faculty at Göttingen were complete, Hilbert somehow found himself at a banquet, seated next to the Nazi minister of culture, Bernhard Rust.

Rust was a wild-eyed super-Nazi, considered mentally unstable even by some others in the government. He added to this perception by continually enacting and rescinding strange decrees rearranging the minutiae of the school week, altering German spelling, requiring students and teachers to greet each other with the Nazi salute, and imposing more decrees along these lines. (Rust shot himself the day after the surrender of the Nazis.) Rust was very personally active in drawing up lists of Jews for purging. A well-known proponent of the "Jewish science" concept, he asserted, "The problems of science do not present themselves in the same way to all men. The Negro or the Jew will view the same world in a different light from the German investigator."[25]

The minister must have heard about the opinions of some in the scientific community about the government's new personnel policies. He turned to Hilbert and inquired if it were true that "the Mathematical Institute *really* suffered so much because of the departure of the Jews?" The brutality of the Nazi regime, which had intimidated so many of his colleagues into silent acquiescence, had not inspired in Hilbert any newfound reticence or concern for his own safety. He spat out his reply.

"Suffered? It doesn't exist any longer, does it!"

6

Emmy Noether in America

"Mathematicians, like cows in the dark, all look alike to me."

The Job Search

Emmy Noether's friends started to search for a position for her abroad as soon as the purges began. When Richard Courant was kicked out and Hermann Weyl assumed the directorship of the mathematics institute at Göttingen, Weyl got in touch with Princeton right away and lobbied for some kind of visiting faculty position for her.[1]

Two crucial American organizations that helped rescue German refugees were the Rockefeller Foundation and the Emergency Committee in Aid of Displaced Foreign Scholars. The Emergency Committee was supported by grants from the foundation.[2]

Many scholars submitted testimonials in support of Noether. One well-respected mathematician wrote to the Rockefeller Foundation and said that, aside from Hilbert, "Miss Noether is unquestionably the most important teacher in Germany."[3]

However, because of the increasing numbers of displaced Jewish academics from Germany and the poor economic conditions throughout the world, it was difficult to find positions for all the people who needed help, including Emmy Noether.

Solomon Lefschetz, a leading Princeton University mathematician who had met Noether while visiting Göttingen in 1931, noted, "Were it not for her race, her sex, and her liberal political opinions (they are mild), she would have held a first-rate professorship in Germany. . . . She is the outstanding refugee German mathematician brought to these shores, and if nothing is done for her, it will be a true scandal."[4] Norbert Wiener of the Massachusetts Institute of Technology wrote, "She is one of the ten or twelve leading mathematicians of the present generation in the entire world. . . . Of all the cases of German refugees, whether in this country or elsewhere, that of Miss Noether is without doubt the first to be considered."[5]

However, when this same Lefschetz was charged with making recommendations to the Emergency Committee, he did not recommend Noether for placement at any prestigious university or research institute. In fact, his recommendation to the committee contained this comment: "It occurred to me that it would be a fine thing to have her attached to Bryn Mawr in a position which would *compete with no one* and would be created ad hoc; the most distinguished feminine mathematician connected with the most distinguished feminine university."[6] (The emphasis in the quotation is very much mine.)

It is painfully clear that Lefschetz, while extravagantly praising Noether's importance as a research mathematician and noting how the backward Germans had denied her an academic position commensurate with her abilities partly because she was a woman, sets about to do the very same thing himself. Lefschetz had another overriding concern while securing some form of appropriate employment for Noether. He wanted to make sure she not be placed in a position to compete with, and quite likely overshadow, his male colleagues. The Emergency Committee followed his recommendation, and so Noether was consigned to what was then, and still is today, a prestigious and high-quality, although somewhat insular, school for women.

Lefschetz's machinations call to mind the discomfort of the young men in the presence of a superior female intellect described at the beginning of the book. It seems plausible that he was eager to spare his colleagues a similar discomfort, of being confronted too regularly with a woman with whom they could not dance.

Did similar considerations haunt Noether's career in Germany? Was the real concern among the academic powers of the time not that a woman could not be the equal of men but the trepidation that some women might even outshine them?

It might be thought that Albert Einstein, already at the Institute for Advanced Study and an advocate for Noether, could have had some pull. However, his influence was limited. The director of the institute, Abraham Flexner, was moved to write the following in a letter:

> Last night Professor Lefschetz, who holds the highest professorship in mathematics in Princeton University and is himself a Russian Jew, came to see me and asked me if I could not in some way shut Einstein up, that he was doing the Jewish cause in Germany nothing but harm and that he is also seriously damaging his own reputation as a scientist. . . . I am seriously concerned as to whether it is going to be possible to keep him and his wife in this country. I have been pleading with them all summer to show the elements of common sense, and their replies have been vain and foolish beyond belief. . . . He and his wife are better taken care of today than they have ever been in their life if they will only behave themselves.[7]

Bryn Mawr

Noether's departure from Germany was delayed because of problems with her visa.[8] However, by October 1933, the issue was straightened out and she was aboard the *Bremen*, sailing for the United States.[9] The manifest described her as "led" (for *ledig*, unmarried), fifty-one, with Göttingen as her place of residence, "Professor" as her profession, and Bryn Mawr as her destination.[10]

Noether had connections in the United States before she arrived there. She had met students and researchers from America when they put in time at Göttingen, and she was, unsurprisingly, quite well known in the States through her reputation as both a teacher and a practitioner of mathematics.[11]

The Bryn Mawr of 1935 was still a young college, having been founded in 1885.[12] It was one of the first institutions in the United States created with a serious educational mission for women. Although it was and is a "college," it had several small graduate programs at its founding. These programs have expanded today to encompass mathematics and several sciences and humanities disciplines.[13] Bryn Mawr was then the only women's college with PhD programs, although women could also pursue graduate work at several coeducational universities by the turn of the century. By 1900, Bryn Mawr had awarded a total of fifteen doctorates, 7 percent of the total awarded to female scholars in the United States.[14] The academic job prospects for these PhD holders, however, were not much better than in Germany. Bryn Mawr was one of the few institutions providing teaching and research opportunities for women commensurate with their qualifications.

Correspondence indicates that the first time Noether learned about Bryn Mawr's existence was when she was told a position had been found for her there. While still at Göttingen, she wrote to her friend and fellow researcher Richard Brauer in September 1933: "Bryn-Mawr is a women's college, but Mitchell and others are there as professors; besides, Veblen has written to me that it is so close to Princeton that they hope I will come over frequently."[15]

The college quickly achieved excellence in mathematics out of proportion to its size. When the *Transactions* of the American Mathematical Society was launched in 1900 in response to the growing mathematical activity in the United States, Bryn Mawr was listed on its title page, alongside Harvard, Yale, Princeton, and others, as a supporting institution.[16] One of Bryn Mawr's founding professors, Charlotte Angas Scott, would become an internationally known mathematician. She became the vice president of the American Mathematical Society and was featured in the first edition of James McKeen Cattell's *American Men [sic] of Science*. So when Noether arrived at Bryn Mawr in 1933, although she knew nothing of the college, the small school had been punching far above its weight in mathematics since its founding a half century earlier.

Bryn Mawr also had some historical connections with Göttingen. Its first graduate fellow in mathematics, who arrived at Bryn Mawr the year of its founding, had studied for a while in Göttingen.[17] Its early professors prepared the curriculum from a variety of texts and recent research papers, including works by Klein and by both Emmy Noether's father and her thesis adviser. Another Bryn Mawr graduate student spent 1894 and 1895 studying at Göttingen. Anna Pell Wheeler, who became an influential American mathematician and the head of Bryn Mawr's mathematics department, had studied there with David Hilbert. Wheeler was in charge of Bryn Mawr mathematics when Noether arrived. Klein's daughter had also studied mathematics and physics at Bryn Mawr.[18]

The Göttingen connections were not completely the result of chance. Even before Noether's rise to prominence at the venerable German institution, it had enjoyed a reputation as at least a bit more welcoming than most to women and students of all races and nationalities. This was particularly true of the mathematics institute, thanks to the egalitarian attitudes of Klein and Hilbert. In fact, around the turn of the (twentieth) century, about half of the most active US female mathematicians had found their way to Göttingen for a portion of their studies.[19]

Bryn Mawr's reputation for excellence in mathematics endures. In 2012, the American Mathematical Society presented its seventh annual award for "an Exemplary Program or Achievement in a Mathematics Department" to Bryn Mawr. In the award announcement, the society provided a fanciful account of how Noether wound up at the storied women's college: "The math department at Bryn Mawr has a long history of encouraging women to pursue careers in mathematics and, more generally, of providing women with an environment in which they may thrive in this pursuit. Indeed, it was Bryn Mawr that Emmy Noether chose as her new home when she was forced to leave her position in Germany in 1933, a choice that is emblematic of the role that the department played in those difficult days."[20]

Of course, Noether knew nothing of Bryn Mawr before it had been selected for her. And despite what turned out to be a brief but joyful experience at the American college, she would undoubtedly have chosen to be situated with her peers at Princeton University or the new Institute for Advanced Study, had she been given a choice.

At the Bryn Mawr convocation opening the academic year of 1933, the college president, Marion Park, announced the arrival of a distinguished foreign visitor to join the faculty:

> I heard yesterday that we are to have a most distinguished foreign visitor ... Dr. Emmy Noether, a member of the mathematical faculty of the University of Göttingen. Dr. Noether is the most eminent woman in mathematics in Europe and has had more students at Göttingen than anyone else in the department. . . . Dr. Noether does not, I understand, speak English well enough to conduct a seminar at once but she will be available for consultation by the graduate students and later I trust can herself give a course. . . . I am glad also

that the college can entertain so distinguished a woman and that the students in mathematics can profit by her brilliant teaching.[21]

When she did finally appear at Bryn Mawr near the beginning of November, her hosts were relieved to discover that her English was, although rocky, better than they had expected.

Although she could communicate effectively, there were things she was not free to discuss, and President Park was protective about this aspect of her situation. Noether had family and friends still living under the Nazis, and she still harbored some hope of one day returning to Germany. At the time, the full unfolding extent of the Nazi horror was not widely known, and its monstrous future could not have been guessed at. For her own sake and the safety of others, Noether could not talk politics.

Bryn Mawr's administration was clearly happy to have a scholar with such a wide international reputation on the faculty. When they discovered that her English was workable, they almost immediately set out to show her off.[22] They arranged for her to give a lecture to which they sent invitations to mathematicians at various institutions in the region.

Noether's early impressions of the populace in her new home seem familiarly European. From one of her letters not long after she arrived: "The people here are all very accommodating and have a natural warmth that's truly winning, even if it doesn't go very deep. You are constantly getting invitations."[23]

The college quickly established the Emmy Noether Fellowship to support researchers to travel to Bryn Mawr and work with the new professor. According to her student Marguerite Lehr, the fellowship supported three research fellows.[24]

Noether was clearly happy at Bryn Mawr, but Germany still occupied the place of home in her mind. It tugged at her. She still entertained the idea of returning to live and work in the country of her birth.

Although antisemitism had been endemic for centuries in most of Europe, she assumed, as did so many of its expatriates and others, that the current extreme state of affairs was an anomaly, that the famous German reverence for scholarship would reassert itself and wash away the tide of thuggery that had temporarily taken power.[25]

Her visit home in the summer after her first partial year at Bryn Mawr showed her the truth. Although she did give a seminar and enjoy some academic contacts there, some of her former colleagues now avoided her, cowering in fear lest someone observe them in conversation with a Jew. She had a school-age nephew named Gottfried who had shown a considerable spark of mathematical talent, whether from nature or nurture, passed down from Great-Uncle Max and Aunt Emmy. But now he told her that he didn't like school, because he had no friends and the other students mocked him for being Jewish, with the support of the teacher.[26] (Gottfried did eventually escape to the United States, where he enjoyed a successful career in statistics, with assistance from an educational fund set up by Weyl to help him.[27])

Germany was no longer an option. Noether had her books and furniture shipped to Pennsylvania and gave up any idea of returning to her homeland.

Professor Noether resumed her academic career at Bryn Mawr with enthusiasm. She soon gathered around her a miniature and female version of the Noether's Boys that were so devoted to her at Göttingen. She brought Göttingen's tradition of peripatetic mathematics to the New World, carrying on animated mathematical discussions with students and colleagues while strolling all over the neighborhood. One of her students, Ruth S. McKee, remembered one such walk:

> One afternoon . . . she was walking with her four students. She liked walking in the country so we started across an open field behind the college. I soon realized that we were heading straight for a rail fence. Miss Noether was immersed

in a mathematical discussion and went merrily along, all of us walking at a good clip. We got closer and closer to the fence. I was apparently the weak sister, concerned mostly in how we would handle the fence. For those of us in our twenties it would be no problem but, from my point of view, however would this "old lady," fiftyish, handle the fence? On we marched right up to the fence and without missing a word in her argument she climbed between the rails and on we went.[28]

During Noether's second year at Bryn Mawr, there was a scramble to secure funding for her position for the future. The school had no vacancy or budget for a permanent faculty position, so the administration sought support from the Emergency Committee, the Rockefeller Foundation, and other organizations. A comment by a Rockefeller official gives an idea of one source of trouble: "Remarkable as are her gifts, there seems every reason to suppose that Dr. Noether would not be able to handle undergraduate work in mathematics. Even the graduate students find her work hard and her standard of what may be expected from them somewhat high."[29]

Park herself may not have helped when, in telling the Emergency Committee that she desired support for Noether for two additional years, the college president commented that Noether was "too eccentric and unadaptable to be taken on permanently at Bryn Mawr."[30]

It seems that Noether's status at Göttingen as a teacher beloved by advanced and talented students but mystifying for the typical undergraduate was redoubled at Bryn Mawr. Then, as now, American college students were less well prepared than their German counterparts.

The same Rockefeller official wrote in his diary:

It has become clear that N. cannot possibly assume ordinary academic duties in this country. She has no interest in undergraduate teaching, has not made very much progress with the language, and is entirely devoted to her research interests. On the other hand the Bryn Mawr authorities like and admire her, and would very much like to have her around for a further period. . . . There is no hope whatsoever of absorption at Bryn Mawr, but there appears to be a fair chance for absorption at the Princeton Institute [for Advanced Study], this being an ideal disposition of the case.[31]

Although the institute had not yet "absorbed" any female members, apparently not everyone in authority shared Professor Lefschetz's concerns surrounding competition.

Despite the misgivings and financial constraints and after much effort, funds were made available for a two-year renewal for Noether's position at Bryn Mawr.

They would never be spent.

Fortunately for all concerned, Noether had little or no contact with undergraduates.[32] During her time at Bryn Mawr, she worked with about a half dozen graduate students, all of whom except one already with doctorates; we might call these postdocs. She supervised one PhD candidate, Ruth Stauffer.

Noether's presence at the school was fairly widely known and appreciated within its community. The *Bryn Mawr Alumnae Bulletin* of February 1935 offered this praise:

During the last and present academic years, Bryn Mawr has been extremely fortunate in having as visiting lecturer an outstanding German mathematician, Professor Emmy Noether. Grants by the Emergency Committee for the Relief of German Scholars and by the Rockefeller

Foundation have made this possible. Prior to 1933, Professor Noether held a professorship at that famous mathematical center, the University of Göttingen. Miss Noether is a leader in the development of the field of Modern Algebra, a subject which made its greatest advances in Germany. The importance and interest of this subject are being recognized in this country and the presence of Miss Noether is of extreme value not only to Bryn Mawr and its vicinity but also to the whole country.[33]

Thomas Underwood, now a teacher of writing at Boston University, interviewed Elizabeth Monroe Boggs in 1979 as part of an oral history project while he was an undergraduate at nearby Haverford College. Boggs's reminiscences, preserved in his notes, are a unique window into some aspects of the Noether experience: "Emmy Noether's arrival had been arranged . . . by the Mathematics Department and they were quite pleased to have her come. Their primary purpose was to give her an environment in which she could continue her creative researches in fairly abstract forms of mathematics, and work somewhat with graduate students. . . . Anna Pell Wheeler, who was the head of the department, allow[ed] me to take a half unit throughout the year of honors work with Emmy Noether."[34]

Standards at Bryn Mawr were high, as Boggs recalled: "I had by that time passed my orals in French and German. The orals in French and German were required of every Bryn Mawr graduate as a demonstration that she could read French and German to the level that permitted some scholarly research. You were not required to speak it."

Boggs added another version to the collection of people's disparate memories about Noether's English. Memory is treacherous: "Emmy Noether . . . spoke no English. That was no problem to Anna Pell Wheeler, who spoke fluent German, but it was a problem for me, because I did not speak fluent German."

At least some people had ideas about the visitor from Germany that would surely have amused her: "I had been warned by Bill Flexner that Emmy Noether was a very lofty personage on whose time I shouldn't expect to trespass, and that continental university traditions about relations between students and professors were such that the professors expected their students to do a good deal on their own and not bother the professors too much, and that all this business of frequent conferences in the professor's office was something I shouldn't expect from her."

At her first meeting with Noether, Boggs got her assignment, which was to study a recently published German textbook called *Modern Algebra*, by B. L. van der Waerden. "This was a book on the abstract forms of algebra," Boggs recalled, "not the kind you get in high school. She told me that I should sit down and study this. . . . I took the book home, and I read enough German so that I could follow along. . . . A lot of it, of course, was in formulas, equations, using symbols which of course looked familiar to me. . . . I could digest two or three pages in an hour. That's the nature of the compactness of the algebraic language. . . . I managed pretty well, and I didn't think it was anything that I needed to ask Emmy about."

After all this independent study, it was time for another meeting with the professor, Boggs recalls: "Come January, Anna Pell Wheeler gently conveyed to Emmy Noether that she had to provide a grade for me, so I was summoned into the presence. . . . She told me that there were some problems in the back of the chapters, and that I should go and pick out some of those problems to do and come back. . . . It was possible for me to do them almost entirely symbolically. I didn't have to write any German or English to do them. . . . Apparently I did them right— particularly when you can choose the ones you're going to do. So I was duly scooted through the first semester with an A."

Elizabeth Boggs went on to become a prominent advocate for people with developmental disabilities.[35] She died at the age of eighty-two in a car accident as she was driving on an icy road.

The textbook assigned to Boggs bears the subtitle *Based on Lectures by E. Artin and E. Noether.* Van der Waerden, the book's author, had attended lectures by both of these mathematicians.[36] He then contributed his own ideas and produced a systematic synthesis of this evolving area of modern mathematics. The only good textbook on modern algebra in the world, it was soon translated into (at least) English and, for decades after its appearance in 1930, was the standard text in the field.[37]

MODERNE ALGEBRA

VON

Dr. B. L. van der WAERDEN
O. PROFESSOR AN DER UNIVERSITÄT
GRONINGEN

UNTER BENUTZUNG VON VORLESUNGEN
VON
E. ARTIN und E. NOETHER

ERSTER TEIL

BERLIN
VERLAG VON JULIUS SPRINGER
1930

Of its author, Noether's friend and fellow mathematician Pavel Alexandrov would later say that he was "one of the brightest young mathematical talents of Europe. Van der Waerden quickly mastered Emmy Noether's theories, added significant new results, and, more than anyone else, helped to make her ideas widely known."[38]

According to Weyl, van der Waerden had arrived in Göttingen from Holland "as a more or less finished mathematician and with ideas of his own; but he learned from Emmy Noether the apparatus of notions and the kind of thinking that permitted him to formulate his ideas and to solve his problems."[39]

It was mainly the promulgation of algebra that occupied Noether's time at Bryn Mawr. This subject is not what we learn as algebra in school. That bane of young people's existence is a set of mainly mechanical techniques aimed at solving for unknown variables in equations or systems of equations. We learn such favorites as the quadratic formula and tricks for factoring polynomials.

What mathematicians call modern algebra, abstract algebra, or simply, algebra is something else entirely (although it, like so much else in mathematics, is historically connected with the solution of equations). Algebra is about what mathematics is always about: the search for pattern, for structure, for the hidden connections among ideas. It is, above all, about abstraction.

One of several structures that form part of algebra is the *group*. A group is the structure that mathematicians use to express and study symmetry, and for this reason, the idea pops up periodically in the history of physics. Group theory is one of the essential tools that Noether drew upon to prove her theorem. A group is essentially is a collection of *elements* and an *operation* that combines them, along with a small set of rules. An example of a group related to symmetry is a set of angles though which we can rotate a particular object or shape; the operation combines two angles. For example, a rotation of 10 degrees combined with one of 35 degrees produces a rotation of 45 degrees. All of the symmetries that appear in Noether's theorem can be described by a particular class of groups called a Lie group (named after Norwegian mathematician Sophus Lie).

Groups had been studied since before Noether was born. But abstract algebra, of which they are a part, has grown continuously up to the present and will remain a central part of mathematics forever. Noether is a major figure in this development. As van der Waerden points out, a large part of his textbook is an explanation of her ideas.[40] Her fame among mathematicians is mainly due to her many results in algebra; in fact, she is considered to have played the major role in defining the field in its modern form. As usual with Noether, this influence went beyond her formally published results. Perhaps even more influential were her methods of working and thinking about mathematics—approaches that she diffused throughout the culture by her teaching, lecturing, and constant communications with mathematicians around the world.

Historically, group theory found its major physics application in crystallography. This use of the theory should seem natural, as the study of patterns and symmetries in space is central to that field. As I'll talk about in the next chapter, group theory became important again in quantum mechanics and particle physics, where, after gravitation, Noether's ideas had their most important influence.

Abstract algebra, including the theory of groups, is what Noether was teaching at Bryn Mawr. Her mission, aside from her own survival and that of her associates and family back in Germany, was to bring modern algebra to America, where it was still relatively unknown. Most of this material, now part of the standard education of any math student, was pretty new stuff at the time.

Near the beginning of her second year, Noether began to travel to New Jersey every week to give a lecture at the Institute for Advanced Study.[41] She sometimes made the three-hour trip by train and was at other times driven there by Wheeler.[42] Noether didn't set foot in Princeton University proper, which, as she mentioned in one of several surviving letters from Bryn Mawr, "admits nothing female."

She was probably not aware that Princeton also admitted very little that was Jewish.[43] The university used a secret quota system that it only began to give up after the end of World War II, replacing it in more recent years by quotas limiting the enrollment of Asian students, justified using similar language.[44]

The Institute for Advanced Study was a unique institution. It had its origins in a suggestion made by Oswald Veblen, a mathematician and correspondent with Noether, to the Rockefeller Foundation.[45] Abraham Flexner, retired from the foundation, made it happen, and Veblen became its professor number one. Veblen, in turn, recruited Einstein to be the institute's professor number two.

Hilbert's famous student Richard Courant seems to have transported some of Göttingen's spirit of ambulatory mathematics to the United States, to which he had escaped from the Nazis in 1934. One American student described his encounters with Courant: "It was so darn human. I had always thought of mathematics as something where you sit down at a desk and apply yourself. But now I found out you could do it while cutting grass."[46]

The institute was to be a place where the smartest people in math and science would be given offices, a generous salary, and no responsibilities. It seems to have worked. However, the concept was not without its detractors. The American physicist Richard Feynman could see the dark side:

> When I was at Princeton in the 1940s I could see what happened to those great minds at the Institute for Advanced Study, who had been specially selected for their tremendous brains and were now given this opportunity to sit in this lovely house by the woods there, with no classes to teach, with no obligations whatsoever. These poor bastards could

now sit and think clearly all by themselves, OK? So they don't get any ideas for a while: They have every opportunity to do something, and they are not getting any ideas. I believe that in a situation like this a kind of guilt or depression worms inside of you, and you begin to worry about not getting any ideas. And nothing happens. Still no ideas come. . . . I find that teaching and the students keep life going, and I would never accept any position in which somebody has invented a happy situation for me where I don't have to teach. Never.[47]

One of the institute's particular features was a large woods. Veblen himself accumulated the land in the name of the institute, bargaining for it plot by plot. He aspired to provide a serene environment and to erect some insulation against the ordinary world beyond. These woods no doubt helped make some of the institute's denizens, and Noether when she visited them, more at home, as the environment must have at least partly replicated the Göttingen forest that served as the venue for so many mathematical strolls and that still surrounds Göttingen University today.

These "tremendous brains" would eventually be housed in an elegant building containing a wood-paneled library, study rooms, and offices with fireplaces. The building even had a locker room with showers and baths to make it more convenient to take advantage of the nearby tennis courts. The faculty song contained the lines:

> *He's built a country-club for Math*
> *Where you can even take a bath.*

Flexner did not want to involve himself in recruiting scientists for the institute. He told Veblen that, to him, "mathematicians, like cows in the dark, all look alike to me." Nevertheless, after things got going, his management style became so offensive that Einstein led the faculty in a mutiny against him.

Noether's lectures at the institute were part of her Bryn Mawr mission to promote the cutting edge of modern algebra, mainly honed by herself, and her uncompromisingly abstract methods to any mathematicians within earshot, whether US born or immigrants such as herself. She was keenly aware that her methods were still strange to some in her audience, many of whom had not been exposed to the Göttingen mathematical culture: "I'm beginning to realize that I must be careful; after all, they are essentially used to explicit computation and I have already driven some of them away with my approach!"[48]

These regular lectures sometimes turned into seminars on specific topics that would span some weeks. Always the teacher, Noether would often bring her postdoc Olga Taussky with her, tasking her with providing lectures at the institute. She would have Taussky rehearse the talks in front of the other students at Bryn Mawr on Mondays to prepare for the (usually) Tuesday lectures at Princeton.

Taussky had earned a PhD from the University of Vienna and had visited Göttingen in 1931 at the invitation of Courant to help edit Hilbert's collected works, when she was only twenty-five. (Hilbert himself would have nothing to do with preparing a collected works. He had no interest in looking to the past and had a distaste for such projects.) While at Göttingen, she began attending Noether's lectures, and the two became friends.[49] Taussky would go on to publish over three hundred papers.[50] To study with Noether, Taussky postponed accepting a grant to work at Cambridge University; she was supported instead by the Noether Fellowship while at Bryn Mawr.[51]

Relations between Noether and Taussky had been periodically dramatic but always ultimately positive. Taussky related how she had "blown up" at Noether when the professor had criticized, in a lecture, a published proof in a way that Taussky thought was unfair.[52] Everyone, including Taussky, sank into their seats and waited for the fallout. Taussky did not yet know her teacher well. Of course, Noether was unperturbed. Taussky recalled, "I recognized for the first time that she was a person who did not mind criticism."

It seems that Olga Taussky had imported to the United States her tendency to be occasionally irritated with Noether. During the Bryn Mawr interval, there was some lighthearted friction between the two women. Taussky would complain about Noether's criticism of her, as the professor would tease her about her accent and clothing.[53] The younger woman was also somewhat offended by some of the attitudes of her teacher, who was a generation older. According to Taussky, Noether had confessed that she actually gave preference to men when recommending people for positions so that they could start families. Taussky also commented that "she was very naïve and knew very little about life. She saw women as being protected by their families."[54] (Noether also made unsuccessful efforts to set her students up with husbands.[55])

Whether Taussky's claim is true or not, it shouldn't be utterly surprising. None of us can see completely around the blinders affixed to us by our time and place. None of us has the moral genius to place our actions within a crystalline sphere, isolated from the prejudices of our environment, where they will survive the judgments of any more (or less) enlightened future society. We are even insensible to our own visitation upon others of the injustices inflicted upon ourselves.

As I explained in Chapter 4, although Hilbert is sometimes described today as a campaigner for women's rights because of his demand that Noether be treated the same as men in her position, he was in fact campaigning on behalf of science. Science would benefit from meritocracy; sex discrimination would harm the progress of mathematics because it would eliminate the contributions of a talented researcher.

Noether's purported recommendation policy tells us something about her attitude and thought process. In following it, she made a utilitarian calculation, bestowing the greatest benefit to the largest number of people. From the point of view of twenty-first-century liberal societies, her policy is flawed: we demand a deontological commitment to demographic equality transcending the foreseeable consequences of our decisions. I don't propose to criticize this stance, which I fully share. I merely point out the fragile and transient nature of ethical fashions. Indeed, the recent strengthening of a cluster of regressive attitudes sometimes called identity politics places traditional liberal values in palpable danger.

Noether was not as single-minded as Hilbert. He was playing a chess game where mathematicians were the pieces and winning meant finding truth in mathematics. Advocating for Noether was part of his midgame strategy. It's obvious that Noether cared deeply about helping people in their careers and their personal lives, perhaps almost as much as she cared about mathematics, and she had her own strategies for optimizing other people's happiness. Sometimes, that strategy took the form of letting someone else take credit for one of her results. At other times (perhaps), it took the form of recommending a man over an equally talented woman. Neither Noether nor Hilbert were egalitarian by modern standards.

Emmy Noether had friends and associates ensconced at the institute, including Hermann Weyl. Weyl himself was not Jewish, although, as noted earlier, his wife was; this made his children Jewish in the eyes of German law. Because of this, remaining in Germany would have been hazardous for him and certainly for his family. After delaying for an alarming amount of time, he finally agreed to come with his wife and two boys to New Jersey.

It would be ungracious to be specific, but, in looking over the roster of names in mathematics and physics absorbed into the institute around this time, one notices, in addition to obvious choices such as Einstein and Weyl, others who were clearly not in the same league as Noether. By then, her reputation as one of the most important mathematicians in the world had been firmly established; in contrast, some of the men at Princeton were comparatively minor figures. There is little doubt that a male Emmy Noether would have been given an office in the same building where Einstein was situated.

During her time at Bryn Mawr, Noether was also busy visiting various expatriates of her acquaintance who had settled in the region, and she attended mathematical conferences in the United States and elsewhere. She also went on sightseeing excursions in Wheeler's car. One such trip to a conference at Columbia University included a leg on the New York City subway, with Noether and her associates chattering animatedly about mathematics the whole time, oblivious to their surroundings.[56]

Her living arrangements while she was teaching at Bryn Mawr were described by a former student as "modest but comfortable." She lived in a room and board close to the campus and was looked after maternally by the owner, a Mrs. Hicks.[57]

Noether and her students often hung out with Wheeler, either having tea, which the math department head enjoyed serving in her apartment, or going on birding expeditions in the Bryn Mawr woods.[58]

Although she taught in English, the books and papers that Noether used for her classes were sometimes in German. She was not always clear which English words should be used for some of the technical terms found in these texts. The upshot was that the mathematical discussion between Noether and her students at Bryn Mawr was conducted in a kind of English-German mélange.[59]

Noether's style in teaching her small group of advanced students at Bryn Mawr was very much one of mutual discovery rather than lecturing. Those of her students who have commented on their experiences with her remember Noether as uniquely effective in bringing out their talents—as a "great teacher."[60]

In 1934, Noether and Weyl created the German Mathematicians' Relief Fund to organize financial support for struggling German immigrants.

With some success, they solicited pledges from expatriate Germans who had been lucky enough to find jobs in the United States and elsewhere. These donors pledged a small percentage of their income to the cause. Noether was also very active in helping anyone still in Germany affected by the growing oppression of the Jews, directly or indirectly, and others who needed help in finding employment inside Germany or in escaping.

One such person was her brother Fritz, who was still in Germany as a professor at the Breslau Institute of Technology. He had been spared, for a while, removal from his position because of an exemption extended in German law to decorated war veterans.[61] But before long, the regime found a way around its own laws and Fritz was threatened with the purge. As had happened with Edmund Landau at Göttingen, Fritz became a victim of mobs of protesting students secure in their own righteousness, the timeless characteristic of such mobs occupying any location on the political spectrum. Also like Landau, Fritz Noether applied for and was granted an early retirement.

Emmy Noether knew that the German pension would not be enough for Fritz and his family, and she was interested in getting him over to the United States. That idea did not present the best prospects, either. Since he wasn't an academic mathematician like her but was of the applied sort, his most likely source of employment would be in industry. Because of the economic depression, there was not much work available in the private sector. Nevertheless, she tried for an American academic post, having heard that the Massachusetts Institute of Technology physics department, the University of Michigan, and some others might be interested. She talked to many people and wrote many letters.

Things were going better for Emmy Noether than they had at any other stage of her life. She delighted in her students, America, her continued research, and her stimulating association with her brilliant colleagues at the Institute for Advanced Study. With her growing affection for America, she let go of the notion of eventually settling in the Soviet

Union—an idea that had remained in her imagination even after she had moved to Bryn Mawr. She wrote that "staying here has the great advantage that—despite the dollar valuation—one can travel almost anywhere. . . . It's also a fact that English seems to have devoured all my memories of the Russian language."[62] In another letter she reiterated, "By the way, my English is absolutely smooth, with the result that the last miserable remains of my Russian have disappeared."[63]

From the perspective of this moment in history, anyone's interest in voluntarily moving to the USSR may seem odd. But in the 1930s, the thuggery of the Bolsheviks and the murderous psychopathy of Vladimir Lenin and Joseph Stalin were still not widely known outside the sphere of their victims. The intelligentsia of Western Europe and America, especially those, such as Emmy Noether, with lifelong socialist sympathies, often viewed the Soviet experiment as at least an interesting alternative to the governmental and social systems under which they lived.[64] And especially between the wars, these systems admittedly left much to be desired. While there was plenty in the way of warning arriving on the lips of those who managed to escape the Soviet prison state, these voices tended to be drowned out by a combination of propaganda and the human tendency to ignore inconvenient information.

Noether had especially warm feelings toward Russia, as her year there as a visiting professor was marked by fruitful collaborations and a strong receptivity to her teaching and ideas, many of which were new to the Russian mathematicians.[65] She had liked what she saw of Russia. All things considered, Noether's harboring of a desire to return there is not surprising.

However, she decided to stay put at Bryn Mawr. She confessed to a friend that her time in the United States had been the happiest in her entire life, because she was valued in Bryn Mawr and Princeton as she had never been in her own country.[66]

Shortly after she made this remark, the underlying cause of a marked external swelling created enough distress that she finally sought medical help. The doctors discovered that she had a uterine cyst of alarming size.

On April 10, Noether checked into Bryn Mawr Hospital for surgery. All went quite well. Visiting her in the hospital after the operation, her friends found her cheerful as always, expressing pride in the fact that only the cyst needed to be taken out and that everything else was intact. The operation had been routine.

A short time later, she was stricken with a sudden fever and died.

Her doctors described what happened in letters to President Park.[67] After what appeared to be a normal recovery, her temperature suddenly soared to 108°F on the fourth day. A blood vessel in her brain had burst, leading quickly to her death. They pointed out that, while the surgery might have had some relation to her demise, she would certainly have been done in by the cyst without it.

Although Noether had only had a few students, her sudden death obviously affected them both personally and academically. The interview with Elizabeth Boggs reveals how the student fared after the tragedy. She had continued her largely independent study of van der Waerden's text, and then: "I found myself in the middle of April with graduation a month away, six weeks at the most, with no major professor and no one but myself knowing what I had been doing. . . . I could not discuss what I had been doing with anyone, because I didn't know enough German to say it in German and I didn't have the English words for all that mathematical stuff. I was somewhat nonplussed."

Wheeler came up with a solution. She drafted the late Noether's most advanced student to work with her beginner. "Olga Taussky's English was not much better than Emmy Noether's," Boggs recalled, "and my Czech was nil, but we met, and she did a rather genius thing. She said, 'Here is a theorem which I know is true because it has been proven by non-algebraic methods. No one has ever proved it by using algebraic methods. I would suggest that you do this as an original exercise.'"

It hardly seems fair, assigning a research problem of this nature to an undergraduate, but Boggs pulled through. "She and I met several time[s]," Boggs said, "and I asked her questions, and we had conversations about it. I remember the terrible panic . . . three or four days before this thing had to be in. . . . I had written it all up . . . and then I suddenly realized that I had made a mistake fairly early in the development . . . so I eventually patched it up and it was accepted."

Marguerite Lehr, another of Noether's students, made these remarks at the small memorial service at the Bryn Mawr chapel:

> At the opening Convocation in 1933, President Park announced the coming of . . . Dr. Emmy Noether. Among mathematicians that name always brings a stir of recognition; the group . . . waited with excitement and [made] many plans for Dr. Noether's arrival. . . . There was much discussion and rearrangement of schedules, so that graduate students might be free to . . . consult with Miss Noether until she was ready to offer definitely scheduled courses. . . . It seemed that a slow beginning might have to be made; the graduate students were not trained in Miss Noether's special field—the language might prove a barrier. . . . When she came, all of these barriers were suddenly non-existent, swept away by the amazing vitality of the woman whose fame as the inspiration of countless young workers had reached America long before she did. In a few weeks the class of four graduates was finding that Miss Noether could and would use every minute of time and all the depth of attention that they were willing to give. In this second year her work had become an integral part of the department. . . . Her group of graduates has included three research fellows here on scholarships or fellowships specifically awarded to take full advantage of her presence. . . . The young scholars who surrounded her

in Göttingen said that [they] were called the Noether family, and that when she had to leave Göttingen, she dreamed of building again somewhere what was destroyed there. We realize now with pride and thankfulness that we saw the beginning of a new "Noether family" here. To Miss Noether her work was as inevitable and natural as breathing . . . but that work was only the core of her relation to students. She lived with them and for them in a perfectly unselfconscious way. She looked on the world with direct friendliness and unfeigned interest. . . . Many a Saturday with five or six students she tramped the roads with a fine disregard for bad weather. Mathematical meetings at the University of Pennsylvania, at Princeton, at New York, began to watch for the little group, slowly growing, which always brought something of the freshness and bouyance [*sic*] of its leader. . . . Miss Noether continually delighted her American friends by the avidity with which she gathered information about her American environment. She was proud of the fact that she spoke English from the very first; she wanted to know how things were done in America. . . . She attacked each subject with the disarming candor and vigorous attention which won everyone who knew her.

Emmy Noether might have come to America as a bitter person. . . . She came instead in open friendliness. . . . Our final consolation is that she made here too a place that was hers alone. . . . We feel profoundly thankful for the assurance that her friends have brought us—that her two short years at Bryn Mawr were happy years.[68]

Death is not without its mercies. Emmy Noether was spared the knowledge of what befell her beloved brother Fritz.

Her efforts to find him a job were dead ends. After she was gone, Weyl continued her campaign, also without success. Fritz had crossed the threshold of his sixth decade. He found a job in Tomsk Polytechnic University in Western Siberia and moved there with his family.[69] It wasn't long before life in Siberia had become so unbearable for his wife, Regina, that she attempted suicide. Fritz brought her back to her family's hometown in the Black Forest so that her sisters could look after her while he continued to earn a living in Tomsk. When he returned to visit her the next summer, she had already taken her own life, three months after Emmy Noether's death.

Near the end of 1937, the NKVD, the Soviet secret police that was the predecessor to the KGB, arrested Fritz Noether at his home in Tomsk. He was charged, absurdly, with spying and sabotage. His sons Gottfried and Hermann were told nothing but themselves were given ten days to leave the country. Meanwhile, Fritz was of course found guilty and sentenced to twenty-five years; his property was confiscated. He disappeared into the Soviet penal system.

After tremendous effort on the part of Weyl and others, Fritz's two boys, now stateless, were allowed to enter the United States. Gottfried became an eminent statistician. Hermann had an illustrious career in chemistry.

Both sons got married in America. Gottfried served in the US Army. In fact, he joined many other German speakers in becoming a wartime spy for the Allies and translated at the Nuremberg trials. Meanwhile, the brothers still did not know what had befallen their father.

Some fragmentary information eventually reached them: Fritz had been seen alive in prison in 1939 and 1940 and had been seen in Moscow in 1941 or 1942. When Mikhail Gorbachev came to power in 1985, Hermann Noether wrote to him, and finally, in 1989, the USSR provided an official account.

Fritz Noether had been executed on September 10 or 11, 1941, for anti-Soviet agitation. However, his case had been reviewed in

December 1988 by the Soviet Supreme Court, which ruled that he had been groundlessly arrested and wrongfully convicted.

Emmy Noether's unique character was such that she seemed almost supernaturally immune to the feelings of bitterness and jealousy that would be likely to overwhelm most people under the same circumstances. Her students at Bryn Mawr were, on some level, aware of the comparative injustice in Noether's not having been invited to the prestigious Institute of Advanced Study but instead having been placed (to the students' own benefit) in a respectable women's college but hardly a world-renowned center of scientific activity.[70] They also noted her undented cheerfulness, her enthusiastic engagement in teaching the small band of students entrusted to her. These students, although equally enthusiastic and undoubtedly capable, were unprepared in the abstruse reaches of abstract algebra to which participation in Noether's research would have to carry them and, therefore, needed to struggle to acquire the necessary background.

In the words of her student Ruth McKee, "Those of us who knew Miss Noether will be forever indebted to her for her example of selfless living."[71] In the assessment of physicists Leon Lederman and Christopher Hill, "She was one of the greatest mathematicians in history, yet quiet, ascetic, and kind. Few outside the precincts of mathematics and physics have ever heard of her. Here is a role model for everyone."[72]

In Memoriam

Emmy Noether was cremated in a quiet ceremony attended by a small gathering of friends. Two weeks later, a public memorial service was held at the college. Mathematicians arrived from several countries to pay their respects, many contributing moving and personal eulogies and reminiscences.

BRYN MAWR COLLEGE

INVITES YOU TO BE PRESENT AT A

MEMORIAL SERVICE FOR

EMMY NOETHER

PH. D. UNIVERSITY OF ERLANGEN, 1907

AUSSERORDENTLICHER PROFESSOR OF MATHEMATICS, UNIVERSITY OF GÖTTINGEN, 1922-33

VISITING PROFESSOR OF MATHEMATICS AT BRYN MAWR COLLEGE

BORN MARCH 23, 1882, DIED APRIL 14, 1935

IN GOODHART HALL

ON FRIDAY AFTERNOON, APRIL THE TWENTY-SIXTH

AT THREE O'CLOCK

AN ADDRESS IN APPRECIATION OF HER WORK AS A

SCHOLAR AND TEACHER WILL BE DELIVERED BY

DR. HERMANN WEYL

PROFESSOR OF MATHEMATICS AT THE INSTITUTE FOR ADVANCED STUDY

Hermann Weyl's eulogy has been widely quoted, as it is so moving and contains so many unique impressions and details. Like many others, he experienced the news of Noether's death as a shock: "We thought her to be on the way to convalescence when an unexpected complication led her suddenly on the downward path to her death within a few hours. She was such a paragon of vitality, she stood on the earth so firm and healthy with a certain sturdy humor and courage for life, that nobody was prepared for this eventuality."[73]

Weyl remembers their time together at Göttingen and his attempts to improve Emmy Noether's position: "When I was called permanently to Göttingen in 1930, I earnestly tried to obtain from the Ministerium a better position for her, because I was ashamed to occupy such a preferred position beside her whom I knew to be my superior as a mathematician in many respects. . . . In my Göttingen years . . . she was without doubt the strongest center of mathematical activity there, considering both the fertility of her scientific research program and her influence upon a large circle of pupils."

He admits his defeat in the battle to improve Noether's position, as well as his failure to get her admitted to the Göttingen Academy of Sciences and Humanities.

Weyl notes Noether's unusual timeline. Unlike most mathematicians, who burn bright in their twenties and by the time they reach middle age are more or less reduced to cool embers, Noether's powers built slowly and inexorably and never showed any sign at all of cooling down.

He continues with admiration of her mathematical talent: "She possessed a most vivid imagination . . . she could visualize remote connections; she constantly strove toward unification. . . . She possessed a strong drive toward axiomatic purity."

Weyl winds up with praise of Noether's personal qualities: her sense of humor; her boldness; her frankness; and, above all, her generosity. These remarks then veer off into comments that have struck many modern readers, and strike me, as wildly inappropriate, touching on Noether's appearance and sexuality. There's no doubt that Weyl's heart was in the right place, but it's unfortunate that he didn't have the opportunity to let a friend take a look at his eulogy and chastise him with a "Really, Weyl, you can't say these things."

The eulogy does end well, with phrases that have become some of the most widely quoted appreciations of Emmy Noether: "Her heart knew no malice; she did not believe in evil—indeed it never entered her mind that it could play a role among men. . . . The memory of her work in science and of her personality among her fellows will not soon pass away. She was a great mathematician . . . and a great woman."

After the memorial service at Bryn Mawr, mathematical societies around the world held their own memorials, ranging from moments of silence during regular meetings to special services with the reading of tributes and recollections. The most relevant of these was the commemorative meeting of the Moscow Mathematical Society, held on

September 5, 1935. Her brother Fritz made the trip from Tomsk to attend. In addition to Emmy Noether's friends and people directly associated with her, many mathematicians who were attending the First International Conference on Topology, which was also taking place in Moscow, were there. As is not unusual with mathematicians, many of those who stood spoke mainly of mathematics: the importance of Noether's work to them and their fields. Among the speakers were John von Neumann, André Weil, and Solomon Lefschetz. As a topologist, Lefschetz discussed Noether's crucial work in advancing that field.

After a ceremonial moment of silence, the president of the society and good friend of Noether's, Pavel Alexandrov, spoke: "The death of Emmy Noether is not only a great loss for mathematical science, it is a tragedy in the full sense of the word. The greatest woman mathematician who ever lived died at the very height of her creative powers."

He mentions the importance of Noether's theorem and of Noether's related papers published around the same time, and then describes how "she became the creator of a new direction in algebra and the leading, the most consistent and prominent representative of a certain general mathematical doctrine—all that which is characterized by the words [abstract mathematics]."

He goes on to give his impressions about the importance of Noether's research in algebra: "These ideas [have] already led to many specific fundamental applications. . . . If the development of mathematics today is moving forward under the sign of algebraization . . . then this only became possible after the work of Emmy Noether. It was she who taught us to think in terms of simple and general algebraic concepts . . . and not in terms of cumbersome algebraic computations . . . the influence of Emmy Noether's ideas . . . is also keenly felt in H. Weyl's book [*Group Theory and Quantum Mechanics*]."

He ends by sharing his personal feelings: "With the death of Emmy Noether I lost the acquaintance of one of the most captivating human beings I have ever known. Her extraordinary kindness of heart, alien to

any affectation or insincerity; her cheerfulness and simplicity; her ability to ignore everything that was unimportant in life—created around
her an atmosphere of warmth, peace and good will.... She loved people,
science, life with all the warmth, all the joy, all the selflessness and all the
tenderness of which a deeply feeling heart—and a woman's heart—was
capable."[74]

The most often quoted Noether obituary appeared in the form of
Albert Einstein's letter to the *New York Times*.[75] The *Times* had briefly
noted the death of what appeared to be an obscure Bryn Mawr math
professor. When Einstein read the one-paragraph obituary, he thought
it was woefully inadequate. Some have claimed, without definitive
evidence, that the letter that appeared a few days later was written by
someone else and signed by Einstein. Whether he penned the letter or
not, Einstein surely agreed with its contents. And no matter who the
author is, the letter is eloquent.

> Pure mathematics is, in its way, the poetry of logical ideas.
> . . . In this effort toward logical beauty spiritual formulas
> are discovered necessary for the deeper penetration into
> the laws of nature. . . . Emmy Noether, who, in spite of the
> efforts of the great Göttingen mathematician, Hilbert,
> never reached the academic standing due her in her own
> country, none the less surrounded herself with a group of
> students and investigators at Göttingen, who have already
> become distinguished as teachers and investigators. Her
> unselfish, significant work over a period of many years was
> rewarded by the new rulers of Germany with a dismissal,
> which cost her the means of maintaining her simple life and
> the opportunity to carry on her mathematical studies. Far-
> sighted friends of science in this country were fortunately
> able to make such arrangements at Bryn Mawr College and
> at Princeton that she found in America up to the day of her

death not only colleagues who esteemed her friendship but grateful pupils whose enthusiasm made her last years the happiest and perhaps the most fruitful of her entire career.

Einstein seemed to be correct about "the happiest," in support of which we have Noether's testimony, but we can certainly quibble about "the most fruitful." In truth, she was heavily preoccupied in America with her strenuous efforts to try to save her brother and his family and to help in any way she could her friends and colleagues trapped in the old country. She was burdened, as well, with the uncertainty of her own position and its future financial support. The result was that, although she did pursue research fruitfully and publish, these months in Bryn Mawr were far from her most productive. That she could carry out any original mathematical investigation at all, under the circumstances, is impressive.

Her ability to perform so much creative work under such conditions puts her in that unique category of human whose brains are able to function at the most rarefied levels regardless of the physical and emotional circumstances. The astronomer Karl Schwarzschild is another impressive example. While Schwarzschild was fighting in the trenches with the German army in World War I, he absorbed the general theory of relativity from copies of Einstein's papers.[76] He improved on Einstein's calculation of the precession of Mercury's orbit, one-upping Einstein's approximate calculation with an exact derivation. That achievement, however, was just a warm-up. He then sent a letter to Einstein from his post (the Germans had good mail service in their, it must be said, comparatively luxurious trenches) that astonished the recipient. Schwarzschild had found the first exact solution to Einstein's gravitational equations, something that Einstein didn't even know was possible. The astronomer had made a routine of doing physics calculations in between firing cannons at the Russian troops. (Schwarzschild, who was Jewish, had volunteered for military service, extraordinarily, as a professor in his forties. He signed up as an expression of his conviction that Jews should do such things to demonstrate their patriotism and to overcome antisemitism. He would not survive the war.[77])

Perhaps the most quoted line from Einstein's obituary, which is the most quoted memorial, is this one:

> In the judgment of the most competent living mathematicians, Fräulein Noether was the most significant creative mathematical genius thus far produced since the higher education of women began.[78]

Einstein's comment has produced a certain amount of puzzlement, partly because of its awkward phrasing. I'm not sure what date one could attach to the moment when "the higher education of women began." In any case, none of the debates about what exactly Einstein had in mind have noticed that he doesn't describe Noether as the reputed most significant *female* creative mathematician. The literal meaning of his statement is this: from sometime perhaps in the late nineteenth century or the early twentieth, when women were allowed, let us say, to matriculate in most universities in most of Europe until the day that he wrote his letter, Noether was *the* top math genius in the world, without qualification. Perhaps that's not what he had intended to say. In any case, Einstein's lovely obituary in the *Times* probably helped make some portion of the public outside the circles of mathematicians and mathematical physicists aware of her existence.

———

The president of Bryn Mawr College immediately turned Noether's death into the springboard for a fundraising campaign. Park and her colleagues expertly plucked the heartstrings of various wealthy people who they suspected might be moved by the tragedy. Their internal correspondence shows them conferring over which potential donors were

---

EMMY NOETHER MEMORIAL FUND

Date..................................

I promise to give for the Emmy Noether scholarship fund to be used for advanced work in mathematics under the direction of Bryn Mawr College:

$.................... herewith

$.................... $.................... on or before May 1, 1936
Total amount of Gift

Name..

Address..

To insure exemption from the income tax for this gift please make cheques payable to **The Trustees of Bryn Mawr College,** mark **For the Emmy Noether Memorial Fund** and mail to Dr. Florence R. Sabin, The Rockefeller Institute, 66th Street and York Avenue, New York City.

---

known to be Jewish and how much to emphasize Noether's ethnic heritage when hitting up those who were.

Florence Sabin headed the committee to establish a memorial fund in Noether's name.[79] Sabin was a biologist who had done important research in anatomy. She had joined the Rockefeller Institute for Medical Research in 1923 at the invitation of Simon Flexner, its director, becoming the first woman with full membership.

Some notes that have survived from a strategy meeting for fundraising include remarks such as "Dr. Simon Flexner's name is the freshest on any Jewish list and is certain to rouse the attention and interest of any one whom he addresses" and "all letters to Jews should mention specifically that Dr. Noether was Jewish and a refugee."[80]

The cultivation of Lucius Nathan Littauer by those raising money for the memorial fund offers a glimpse into the sometimes-subtle machinations of fundraising. Littauer was on the fund's radar because he was one of the most prominent Jewish philanthropists in the country, and he had already promised a thousand dollars toward the renewal of Noether's position.[81]

Littauer was a colorful character.[82] After inheriting his father's glove manufacturing business, he made a huge success of it and amassed an even greater fortune from further ventures. He served in the US House of Representatives and was a close friend of Theodore Roosevelt, who was his roommate at Harvard. Littauer waited years to marry the love of his life, to spare his parents' feelings, for she was not Jewish. He was accused, probably unfairly, of illegally contracting with the government when he was a member of Congress. In 1914, he and his brother were convicted of evading import duties on expensive jewelry.[83] The judge commented from the bench, "For an ex-congressman so far to forget his oath taken five times and knowing so well the provisions of the law he helped to frame seems to be incomprehensible." The article from which I gathered most of the facts about Littauer's life, although otherwise detailed, omits any mention of this incident.

Among his charitable deeds, Littauer funded hospitals, Jewish community centers, and libraries of Hebrew books, and he gave two million dollars to Harvard University. He would endow Bryn Mawr with the Littauer Judaic Book Fund.[84] A statue of Littauer was erected in his home town of Gloversville, New York, while he was still alive. He died in 1944.

The campaign to extract a donation began with a letter from Sabin to President Park on March 21, 1936: "Judge Mack told me that you would have immense influence with Mr. Littauer because he remembered your father so keenly. Perhaps you know that Mr. Littauer used to go with your father to church because there was no synagogue in the town where they lived."[85] A few days later Park wrote to Littauer, described the memorial fund, gently suggested the idea of a donation, and said, "I should like to talk with you unofficially, too, about the early Gloversville of my youth, into whose history your family is so closely built!" Things were looking good just five days later, when Littauer replied, agreeing to see Park and ending with, "It would be an exceptional pleasure to meet you personally and discuss the many incidents of our Gloversville experience."

The gambit seems to have worked. On April 23, Park wrote to thank Littauer for his $1,500 donation. It may not sound like much, even in 1936 dollars (it's equivalent to about $32,000 today), but Littauer was fairly tapped out after his other huge donations. President Park's report to Sabin contained a curious observation: "I gathered somewhat from Mr. Littauer's tone that he was not particularly eager to give to objects with Jewish connection. That is, contrary to Mr. Billikopf's suggestion, he was giving rather because of Dr. Noether's eminence than because she was a Jewish refugee." And, in Sabin's reply: "I am sure that Dr. Flexner does not wish to stress the Jewish question but only the fact of Emmy Noether's genius as a woman." So it seems as if they almost overplayed their hand.

After a final check for $50 from the actress Helen Hayes, the fund was deemed fully funded at the end of November with a balance of $10,100.[86]

Thirty years passed.

In 1964, World War II was something that old people talked about. The Beatles' appearance on the *Ed Sullivan Show* inaugurated the British Invasion. Robert Wilson and Arno Penzias discovered the cosmic microwave background, and the US surgeon general published a report saying that smoking might be hazardous.

In that year, Bryn Mawr moved its math and physics books into a new building. During the operation, someone discovered a mysterious, unlabeled box of ashes in a desk drawer. Somehow, the box came to the attention of a person who remembered: these were Emmy Noether's ashes.

In the three decades since her death, no one had inquired about the disposition of her remains. She had no family in the United States; there was no one to take responsibility.

Bryn Mawr is an extraordinary campus. It's certainly not the only American college or university to have constructed its dormitories and academic buildings in some kind of attempt, perhaps, to create a New World Oxford, but it's the most convincing. It has atmosphere.

The place on campus with the most atmosphere is called the Cloisters, shown in the accompanying photograph. It's most famous for being the site of regular skinny-dipping on the part of Katharine Hepburn.[87] There was another tradition, of a kind: the ashes of three members of the Bryn Mawr community already rested in the ground beneath its paving stones. The de facto custodians of Noether's remains decided that she might as well join them. Two paving stones were removed to expose the earth. Four people gathered in secret and placed her ashes in the ground. It was an early September evening in 1964.

The paving stones were mortared back in place. The graves of the other three, resting under the bricks here and there in the Cloisters, were marked with simple slabs reminding those above who they were and when they lived. But Noether's grave would remain unmarked. There was nothing to indicate that the fresh mortar signified anything but a simple repair job. For fifteen years, passersby strode casually over and around her grave, with nothing to advise them what they were treading upon.

The college then decided to celebrate the centennial of Noether's birth, which would be in 1982. By 1980, the preparations were already underway. Questions were raised: where was she buried?

One person knew. Ethel Whetstone, Bryn Mawr's head science librarian, had been one of the small party gathered around to witness Noether's ashes being tucked under the Cloisters walkway. As far as anyone knew, she was the only one of that group still alive.

Whetstone brought the college archivist to the spot and pointed out the two paving stones. The mortar was still visibly newer there. During the intervening decade and a half after the burial, the librarian had tried to interest the college in showing respect for this grave by installing some type of marker, but she couldn't get anyone's attention.

But now there was to be a public event. It would be a fundraising opportunity. The unseemly state of affairs wouldn't do.

A new memorial stone was laid during a small discreet service attended by a few former colleagues and students. It bears her initials and dates and nothing else. An unprepared visitor who happens to notice the marker will not know who "E. N." is and will tread unawares over the ashes of one of the twentieth century's most important mathematicians and teachers.

Ethel Whetstone retired in 1980. The institutional knowledge of Noether's burial was very nearly lost. Her now-much-photographed gravestone only came to exist through a slender thread of good fortune.

Noether's birthday was celebrated two years later with a symposium of talks by mathematicians, students, and colleagues who had known her.

I was one of those who strolled through the Cloisters, on occasion, during the final few years before Noether's blank stone was replaced by the one bearing her initials. From several states away, I would visit my high school sweetheart at Bryn Mawr whenever my minuscule supplies of time and money permitted it. We both enjoyed the Cloisters atmosphere, how it seemed solemnly to take itself seriously even while unavoidably raising the question of what a cloister was doing there.

This was the time, as a student of physics, when I learned about the content of her great theorem and found it to be the most beautiful thing in science I had ever encountered. But I didn't know about *her*; the result appeared in my textbook of classical mechanics as a disembodied mathematical fact, with no mention of the life behind it—unlike many other results, attached to the names of illustrious men.

I returned to Bryn Mawr for the first time since 1979 to delve into the college's archives for this book. The campus and its vibrations had not changed much. Walking among its assembly of Gothic and modern buildings evoked a mélange of pleasant and melancholy memories. I searched the Cloisters for the gravestone. I had seen pictures of it and knew approximately where it was supposed to be; nevertheless, I missed it the first time around. Even if the cryptic slab had been in place when my undergraduate self had haunted these serene passages, I would have had no idea who "E. N." was.

Noether's theorem slept. From her death in 1935 until the early 1950s, although it was not entirely dead to the world, it slumbered deeply.

7

Reawakening

"How can it be that mathematics . . . is so admirably
adapted to the objects of reality?"

The Standard Model

If the criterion by which we judge physics theories is how accurately
they predict the results of measurements, then the standard model and
general relativity are the two most successful theories in all of physics.

By now I think I can safely assume that my readers have some idea
of the problems that general relativity addresses. The standard model
arose later, beginning in the 1950s. General relativity is a complete solution for gravity, explaining its every aspect and puzzle with a kind of
finality and aesthetic completeness rarely seen in science. However, it
explains nothing else. General relativity is silent about electrical interactions, nuclear forces, and has nothing to say about why we see the elementary particles (electrons, quarks, etc.) that we do and why they have
the masses and other properties that we observe.

All these other aspects of nature are explained by the other great fundamental physics theory of the twentieth century: the *standard model*. This model explains every force and particle in nature, and it predicted the existence of many of these particles and their interactions. It is the modern theory of matter. However, to the standard model, gravity is a complete mystery. Gravity does not fit within its framework.

This is the peculiar status of modern physics: two immensely successful theories, accurate beyond anything seen in science before, each explaining its own area of concern with serene self-confidence, yet utterly silent about each other's domain and at a loss about how to reconcile themselves.

Although the two incommensurate theories deal with separate spheres of phenomena, and although they're drastically different in character and style, they share something that bridges the four-decade span of time between them: Noether's theorem. In fact, the theorem is more important to the existence of the standard model than to general relativity, although the standard model arose long after Noether's death. The theorem was indispensable for making sense of 1915's new theory of gravity and for guiding cosmology and gravitational research thereafter. But general relativity itself, the mathematically correct gravitational field equations that cost Einstein so much debilitating labor, managed to struggle into existence without the help of Noether's theorem (although perhaps not without the help of the theorem's creator).

This was not the case with the standard model. This theory of matter is largely an application of Noether's theorem. Without the theorem, appearing in various special formulations more easily digested by physicists, there could be no standard model. This chapter describes how the theorem's heart was kept softly beating from Noether's death until the 1950s, and briefly traces how its ideas reentered physics and became the basis for the standard model. Since this model, along with general relativity, defines the entirety of our current fundamental understanding of the physical world, we'll see in this chapter just how Noether launched modern physics.

The common claim that the standard model and general relativity are theories of unprecedented accuracy is worth backing up with a few details. Let's admire a few of them now.

The standard model's most impressive accomplishment in this direction was the prediction of something called the *magnetic moment* of the electron to better than three parts in ten trillion (10^{13}).[1] Nothing else about elementary particles has ever been predicted with such stunning accuracy.

Roughly speaking, an object's magnetic moment tells you how eager it is to align with an external magnetic field, just as a compass needle aligns with the earth's field to indicate which way is north. Electrons are extremely simple objects, meaning they have only a few measurable properties. They don't even seem to have any size! But they do have mass, an electrical charge, and something that we call spin, because they have an angular momentum. Physicists know that when something has a charge and spins around, it must also have a magnetic moment, and that's definitely the case with electrons. In fact, all the familiar permanent magnetism around us, from compasses to refrigerator magnets, is the result of the tiny magnetism in billions of electrons adding up to create a big force that we can feel.

The measurement of the electron's magnetic moment to this astonishing degree of accuracy was important for a few reasons. It gives us confidence in the correctness of our fundamental model of matter and force, and it reassures us that our ideas of the nature of the electron are accurate. If, for example, we had an even slightly inaccurate value for its magnetic moment, this result could mean that the electron was some kind of composite particle, with component parts, rather than the simple object that we believe it to be.[2]

From general relativity we get the identity of gravitational and inertial mass discussed in Chapter 2; this mass has recently been measured to the 10^{15} level (about one part in a hundred trillion) with an experiment in a satellite.[3] We also recently witnessed the first direct detections of gravitational waves, a formidable feat of sensitive measurement.

But, as with general relativity, the true value of the standard model is not merely in the calculation of some physical quantity to an impressive number of decimal places, as important as that is in lending confidence to the theory's correctness. Its importance lies in the model's ability to make sense of the universe. The standard model encompasses all of our knowledge of matter, energy, and forces, except for gravity, into a single coherent picture. It's considered by many physicists and others to vie with general relativity as humankind's greatest intellectual achievement.

The standard model was built by combining Noether's theorem with quantum mechanics. The stage was set by Hermann Weyl.

Weyl has already appeared as a benevolent presence multiple times in this book. He was David Hilbert's student, an admirer and supporter of Noether, the author of a book on non-Euclidean geometry, a brilliant traveler spanning the worlds of mathematics and physics, and, most recently, one of the most moving, if at times peculiar, eulogists of the hero of our story.

(Einstein would write later of Weyl that "Weyl is a highly talented and, added to that, a versatile man, very fine and likable as a person as well. We may expect great things from him yet."[4] Of course, Einstein guessed right: Weyl did indeed accomplish many great things, and his name is well known to all students of mathematics and physics.)

Beginning almost immediately after the announcement of the theorem by his colleague and friend, he published, over the following decade, a series of prescient papers that showed how symmetries could lead to particle interactions.[5] Thus Weyl was the first to see how Noether's theorem could have implications beyond space-time symmetries and general relativity and reach into quantum mechanics and the new physics that came with it. In this way, Weyl made up for Noether's relative lack of interest in physics. In the early years of quantum mechanics, he showed how her discovery explained the connection of the new quantum theory with classical physics and how it could be applied to make new predictions in the quantum realm. Around this time, Weyl also expanded on his ideas in a book called *Space-Time-Matter*.[6] There he directly referenced Noether's paper and discussed the theorem's applications.[7] Weyl published this book (in German) in 1918. It was a success, renewed in several subsequent editions and, by 1922, appeared in an English translation. The book was influential and widely cited and would have been familiar to the physicists who would later work on the standard model.

Noether's work in general became known better in America, beginning in the 1920s, through the writings of Oswald Veblen, who published a treatise on differential invariants as a Cambridge Tract in 1927.[8] Veblen corresponded with Noether around this time, but she wrote to him that she had not continued with any research related to (what we now call) Noether's theorem, as she had become absorbed in other areas of mathematics.

Important steps were taken in the 1930s by Werner Heisenberg, the man who injected the famous uncertainty principles into quantum mechanics. This principle explained, for the first time, how nature placed fundamental limits on the precision with which particular pairs of properties could be simultaneously known. Weyl's interest was in teaching the mathematics needed to understand general relativity and to explain how space-time symmetries were related to conservation laws in Einstein's theory, as Noether had shown. Heisenberg considered symmetries related to the internal states of particles, a more abstract type of symmetry akin to the gauge symmetry of electromagnetism described in Chapter 3. In this way, he took a big step in the direction of modern particle physics and the standard model.

During World War II, there was, understandably, a pause in this highly theoretical research. Starting in the 1950s, there was a flurry of activity in particle physics, with important papers coming out every year. Everyone was solving a different piece of the puzzle. It seemed to have come together by 1974, when John Iliopoulos gave a talk titled "Plenary Report on Progress in Gauge Theories" at a high-energy physics conference. His presentation was the first time someone had pulled together all the pieces into a coherent, encompassing model of the fundamental constituents of matter and the forces acting among them; it became known as the standard model.[9]

The gauge theories referred to by Iliopoulos are a class of field theories with a particular type of symmetry. The first example that arose in physics is Maxwell's electromagnetic theory. Recall from Chapter 3 the story of the bird on the high-voltage wire, and why it survives. Because only the *differences* in voltage have a physical effect, we can add a number, a constant, to the voltage and nothing changes. When we can perform a transformation that results in no change, that's the definition of a symmetry. As explained earlier, the type of symmetry exhibited by the electrical potential is called a gauge symmetry, and field theories that exhibit it are gauge theories.

Noether invented the concept of the gauge theory in her 1918 Noether's theorem paper, in an analysis whose later indispensable relevance for physics mathematician-historian Yvette Kosmann-Schwarzbach calls "clairvoyant."[10] Noether's results showed that general relativity was an example of a gauge theory.[11] Gauge theories were to be the model of most of the modern physics to follow, including the standard model of particle physics and most of its proposed extensions and unifications with gravitation, such as string theory.[12]

The paper that is often mentioned as the one that kicked off the standard model, published by Sheldon Glashow in 1961, does not mention Noether's theorem directly. He does nevertheless allude to it: "In the conventional Lagrangian formulation of quantum field theory the relation between symmetries of the Lagrange function and conservation laws is well known."[13]

This remark of Glashow's is highly illuminating. Since Noether's theorem deals specifically with the symmetries of the "Lagrangian," and since their relationships to conservation laws—something that the originator of the standard model says is "well known"—the theorem clearly formed the basis of the standard model from the start. What's less clear is if Glashow and his colleagues were aware of the provenance and history of the result they depended on, or of its full mathematical generality at least at the time of the earliest papers in the field.

In 2020, Glashow looked back over the history of the standard model, which he played such a prominent role in creating, in a technical but useful article published in a science-review magazine of which he happens to be the editor in chief.[14] He initially devotes several paragraphs to a summary of the meaning and importance of Noether's theorem, first in its application to space-time symmetries, with examples from classical mechanics, and then as applied to internal particle symmetries, as initially explored by Heisenberg. The impression is certainly that the builders of the standard model set out with the deliberate program of applying Noether's theorem to the symmetries of particle properties.

Among the key papers of the 1950s, 1960s, and 1970s whose results ultimately became subsumed into the standard model, we can indeed see a ubiquitous application of the Noether program. The papers all employ a similar methodology: the creation, combination, and enlargement of symmetries, and the calculation of the consequences of those symmetries for conservation laws, which sometimes lead to the prediction of new particles. This methodology is what I mean by the *Noether program*.

The papers share another interesting property: rarely does one of them cite Noether's theorem directly, either in a reference or in the text.[15] Almost none of the important papers by the key players in this arena—Glashow, Richard Feynman, Julian Schwinger, Murray Gell-Mann, Wolfgang Pauli, Peter Higgs, and so on—mention Noether's theorem, while making use of it routinely. There are a few interesting exceptions in a handful of other papers, which, although germane, were not part of the main line of development.

In 1972, mathematician and math historian Clark H. Kimberling published an article in which he noticed a recent "surge of interest in Emmy Noether and her mathematics."[16] He notices that despite her apparent importance, there seems to be little mention of her in mathematics history books—something that I may have pointed out in this book here and there. Curious about what he had heard concerning the importance of Noether's work for physics, Kimberling found his way to noted physicist Eugene Wigner, who told him, "Her contribution to physics ... concerns the conservation laws of physics, which she derived in a way which was at that time novel and should have excited physicists more than it did. However, most physicists know little else about her, even though many of us who have a marginal interest in mathematics have read much else by and about her."

Kimberling also remarks that the textbook from which most physicists at the time learned advanced classical mechanics, a book by Herbert Goldstein, uses a version of the theorem extensively without ever mentioning Noether's name.

Here Wigner confirms what my survey of the particle physics literature suggests: many of the physicists working on these problems knew how to apply special cases of Noether's theorem, which thereby became essential in the development of the standard model, but they didn't appreciate the mathematical context of the relationships they were using, and they knew nothing about Noether or her general results.

The later recountings of this period—especially by Wigner, who refers specifically to Noether and her theorem—have the benefit of historical context absorbed later and applied in hindsight. Nevertheless we can be grateful to Wigner for emphasizing the central role of the theorem in the work for which he and several of his colleagues received Nobel Prizes.

Already in 1925, Felix Klein noticed that physicists were not reading Noether's 1918 paper. Even workers in gravitation, for whom Noether's work would have been indispensable, failed to take the paper into account. Hilbert would probably have agreed with Klein's theory for why it had been overlooked: the paper was too hard for the physicists. Writing to Max Planck, Klein said that in Noether's paper, "clear mathematical reasons are given for why the actual conservation laws apply in the case of the special theory of relativity but not in the general theory of relativity. Unfortunately, Miss Noether's work is written very concisely, and its full scope is difficult to grasp on account of the generality of the presentation. This may be the reason why physicists have not read the work."[17]

The version of Noether's theorem used by the particle physicists is not the full-blown version as originally published. Like the simplifications of the theorem used and taught in classical mechanics, it is specialized—in this case, to the problems of quantum field theory. Where did the particle physicists whose work culminated in the standard model learn about Noether's theorem or the version that played such a central role in their program?

There are several routes from Noether's 1918 paper to the flurry of papers of the 1950s and 1960s that used the theorem to invent the standard model. At least some among the small handful of papers listed above that referenced Noether directly would have been seen by everyone working in particle physics. Although the postwar explosion in scientific publishing was already underway, the volume had not yet become unmanageable, as it is today.[18] The days when a physicist could read every paper in *Physical Review* every month and possess an up-to-date mental map of the entire discipline might have been over, but scientists could still keep abreast of publications in their specialties, at least. And any particle physicist would have read any paper published by Wigner, who was a founder of the field and was the man most responsible for introducing a generation of physicists to the use of group theory in quantum mechanics. A paper referring to Noether's 1918 paper in the context of invariance principles, authored by Wigner and two collaborators, appeared in 1965 in the *Reviews of Modern Physics*, a prestigious and widely read journal.[19] Its appearance means that after 1965, at least some particle physicists were likely aware of the connection between a result of Noether's and their well-established method of working.

We can be even more assured of this awareness of Noether's theorem, because of Wigner's Nobel Prize address.[20] Wigner won the Nobel Prize in Physics in 1963, sharing the award with Maria Goeppert Mayer and J. Hans D. Jensen.[21] He received half the total prize, awarded "for his contributions to the theory of the atomic nucleus and the elementary particles, particularly through the discovery and application of fundamental symmetry principles." The other half of the prize recognized "discoveries concerning nuclear shell structure" and was split between Goeppert Mayer and Jensen.

Wigner's talk is fascinating and provocative, at least to a physicist. With simple examples, he elucidates interesting aspects of physical law and returns repeatedly to the fundamental importance of symmetry principles, with the written version of his talk including a footnote reference to Noether's 1918 paper. Wigner places symmetry principles in the foundational position, alongside the laws of nature and the elemental stuff of science, the *event*—the thing that happens: "I would like to discuss the relation between three categories which play a fundamental role in all natural sciences: events, which are the raw materials for the second category, the laws of nature, and symmetry principles, for which I would like to support the thesis that the laws of nature form the raw material."

He mentions the space-time symmetries of classical mechanics and their equivalence to conservation laws, referring to Noether. He then mentions that "in quantum theory, invariance principles permit even further-reaching conclusions than in classical mechanics and, as a matter of fact, my original interest in invariance principles was due to this very fact."

As an aside, Wigner is the author of the famous and often-referenced article "The Unreasonable Effectiveness of Mathematics in the Natural Sciences."[22] The article owes its fame to Wigner's pointing out a mystery that many had not recognized as a mystery but that seems more mysterious the more you contemplate it: just why does mathematics, pursued for its own sake, so often turn out to be perfectly suited for describing nature and solving the problems of physics? It's a puzzle not completely unconnected with the subject of this book. Wigner had a long record of influential rumination on this topic (the peculiar power of mathematics), and his Nobel Prize address is a part of that record. Einstein noticed the same unreasonableness, asking, "How can it be that mathematics, a product of human thought independent of experience, is so admirably adapted to the objects of reality?"[23]

In 1963 an illustrious group of physicist-mathematicians gathered in Grenoble, France, to give summer school lectures under the umbrella title of Relativity, Groups and Topology. One of the courses, Dynamical Theory of Groups and Fields, given by Bryce DeWitt, refers to Noether one time and to her 1918 paper.[24]

Because of the writings of Wigner and others, particle physicists would clearly have been aware of Noether's theorem by the 1960s. But as I've shown, their earlier papers already displayed knowledge of the connections between the types of symmetries treated by Noether's theorem and conservation laws. Considering that the main papers on the standard model from this time don't refer directly to Noether, how might the authors have become aware of the Noetherian methods of working? What were the bridges between Noether, a mathematician working in obscurity in Göttingen during World War I, and the post–World War II generation of physicists scribbling away at CERN, the California Institute of Technology, the University of Copenhagen, and Harvard?

I've described the situation from 1918 until the 1950s as the long sleep of the theorem, but it was not a coma. There was Weyl's *Space-Time-Matter*, which by itself would have been enough to keep Noether's theorem on the physics radar. The particle physicists were a mathematically inclined breed, conversant with gravitation, and likely to have been aware of Weyl's popular monograph.

There was also a textbook called *Methods of Mathematical Physics* by Richard Courant and one David Hilbert.[25] It came out in German in 1924 and went through many editions, appearing in English translation by 1953. The two-volume "Courant and Hilbert," as it's universally nicknamed, was widely used to educate legions of physicists in applied mathematics. By now it has competitors, but the older book is still a standard.

Courant was a student of Hilbert's at Göttingen, arriving there the same year that Noether received her PhD from Erlangen. He would eventually be the main impetus behind the creation of a new Mathematical Institute at Göttingen, with funding from the Rockefeller Foundation. The textbook was mainly written by him, with collaboration from other Göttingen professors, using materials from Hilbert's lectures.[26]

Methods of Mathematical Physics contains a section on Noether's theorem in a chapter about the variational calculus. This treatment in such an influential and widely used text would have been sufficient to maintain a reasonable level of awareness of the theorem within the physics culture.

It may seem puzzling to describe a textbook as influential. Textbooks in mathematics and physics are typically regurgitations of a standard body of well-digested knowledge. They tend to reflect conventional pedagogical approaches and are often, although certainly not always, rather dull.

"Courant and Hilbert" was a different sort of textbook. It was the first of its kind. It explained some recently developed applied mathematical methods at a high level. The book was the only place physicists could see, in print, an assortment of techniques that they could apply in their research problems. Before *Methods of Mathematical Physics*, the only way to learn these things would have been to attend the lectures of teachers such as Hilbert or Klein. Courant's book held the hand of many physicists, both young and established, and its inclusion of Noether's theorem was significant.

In addition to the book, there was Courant's career as an important teacher and his creation of a new teaching and research institute in the New World. Courant had left Göttingen under the looming threat of the Nazis (he was Jewish) in 1934, sailing into New York Harbor on a hot August day.[27] Given a position with a modest salary at New York University, he discovered there a mathematics program in sorry shape.[28] Setting out to remake it in the mold of the institute at Göttingen, he created what would eventually be named the Courant Institute of Mathematical Sciences. Noether's nephew Gottfried taught there for a while. The institute remains an internationally known center of research and teaching. The many physics and mathematics students who studied there were exposed to Courant's ideas of how the inheritance of applied mathematics was to be organized and presented.

Another significant link in the chain joining 1918 and the 1950s was a paper by University of Minnesota physicist E. L. Hill. "Hamilton's Principle and the Conservation Theorems of Mathematical Physics" appeared in the *Reviews of Modern Physics* in 1951.[29] As discussed in Chapter 3, Hamilton's principle refers to a (relatively) modern formulation of classical mechanics where it can be easier to examine symmetry properties—the formulation that played a central role in Noether's development of her theorem. In this paper, Hill explicitly sets out to elucidate Noether's theorem for a wider audience: "Despite the fundamental importance of this theory there seems to be no readily available account of it which is adapted to the needs of the student of mathematical physics, while the original papers are not readily accessible. It is the object of the present discussion to provide a simplified account of the theory."

Hill refers directly to Noether's 1918 paper and related early papers by Klein and others. Hill's paper, in a prominent journal, was widely read and was referenced by important contributors to the standard model. If we include second-order references—papers that cite papers that cite Hill—we see a good sample of key standard model creators. Hill's paper,

appearing as it did just at the beginning of the explosion of particle physics papers, apparently succeeded in its aim of introducing Noether's
methods to a wider audience.

Hill admits that his publication offers up a simplified version. As
Kosmann-Schwarzbach points out in her excellent book on the theorem,
because Hill's paper became the route to Noether's results for many later
physicists, they tended to mistake his restricted treatment for the real
thing.[30] In a largely negative review of another textbook that presents an
eviscerated simulacrum of Noether's theorem, Kosmann-Schwarzbach
further laments that, in this area, physics students "have no luck."[31]

Perhaps. But whatever simplified or specialized versions of the theorem managed to lodge themselves in the minds of postwar particle
physicists were enough to guide their construction of the theory that
explains everything except gravity. Our two grand theories of physical reality have no overlap, and attempts to reconcile them have so far
eluded the efforts of our greatest physicists and mathematicians. But
the scientific mind forever yearns for unity, and therefore the dream of
somehow replacing these two incommensurate descriptions of nature
with a single grand edifice remains, in the minds of much of the current
generation of scientists, the ultimate prize.

And it's not unreasonable to suspect that if the ultimate dream of
theoretical physics ever becomes a reality, and we have a unified theory of all the forces of nature, this theory will rest on symmetry considerations and, again, on Noether's theorem. More than reasonable,
this supposition is likely to be all but inevitable. Noether's theorem is,
on the one hand, indispensable for understanding gravitation, and on
the other, the standard model can be described as an extended application of the result. The theorem has revealed itself to be the fundamental theory-construction kit in the very realms that we now seek to
unify and extend to create the theory of the future. We obviously don't
know what this new theory will look like, but it seems inevitable that
Noether's theorem will at the very least point the way.

That is not to suggest that the theorem and other mathematical considerations alone will be sufficient. Physics, after all, is not mathematics. Hilbert's attempt to unify gravity with electromagnetism continued after he became aware of Noether's theorem, but his line of research was premature and unworkable and never led to any new physics.

Noether had no interest in her result after she had moved on from investigating gravitation, a problem of far more interest to Hilbert than to her. Hilbert himself moved on to other interests almost immediately. The number of scientists working in general relativity, where the theorem's heart continued to beat, was modest; the action now was in atomic and nuclear theory and the new quantum mechanics. Were it not for the important books of Weyl and of Courant and Hilbert, and the later explication by Hill, the theorem, even in its simplified versions, might not have been circulating widely enough to have penetrated the awareness of the particle physicists. The standard model depends so thoroughly on the application of Lagrangian symmetries in the Noether style that it's hard to see how its results might have been reached without some version of the theorem. We would not have modern particle physics without it. Neither would we understand conservation laws in general relativity, without which we would not possess a coherent cosmology. From the smallest scales to the largest, our understanding of physical reality rests to a large extent on Noether's theorem.

In Chapter 2, we saw that general relativity was not, as it's often described, the fruit of a solitary mind but was instead a team effort.[32] However, there is room for disagreement about this interpretation of history. One not-unreasonable point of view is to ascribe sole significant authorship to Einstein, as Hilbert generously wanted to do. Those who do so might claim that the important ideas sprang from Einstein's mind alone and that the massive assistance he desperately sought and received from Grossmann, Noether, and others over a decade amounted to something like clerical assistance with the details of calculations.

As reasonable, or not, as the singular-authorship point of view might seem for general relativity, no one ascribes to a single author model of the other great fundamental theory of modern physics. The standard model is universally recognized as having emerged from the combined struggles of a generation of physicists from all over the world. Its pedigree is partly reflected in its character. Beautiful and powerful as it is, the standard model does not possess the sublime cohesiveness of Einstein's theory of gravitation. General relativity rests on one physical concept, its mathematical form flowing almost inevitably from that idea. It explains one thing: the structure of space-time and its relationship with matter-energy. Gravity itself, as a force, disappears, becoming an aspect of space-time itself.

The standard model, in contrast, is a patchwork. It explains the nongravitational forces using a symmetry formed by combining the separate symmetries of the electrical, weak, and strong forces, and from those symmetries deriving the existence of a small bag of particles. The separate pieces are all still visible. It's a lovely sculpture, but it's made from various bits welded together rather than being carved from a single block of marble.

This is not a criticism. After all, general relativity explains gravity and is powerless to explain anything else. The standard model explains everything else. Its patchwork nature has led another generation of physicists to try to make it more beautiful and satisfying by replacing it with some sort of unified theory, sometimes including gravity as well, in a grand unified theory of everything. The implacable barrier blocking the path from where we are now to any of these visions of unity is the lack of experimental data constraining theory. I'll have a bit more to say about this in Chapter 8.

The standard model may seem to lack unity from a physical point of view, but on a deeper conceptual and mathematical level, it's unified by Noether's theorem. It rests on symmetry and the duality between symmetries and conservation laws explained by Noether in 1918.

This brings us to a cruel irony. The summit of high-energy physics, the standard model, is based on a systematic application of symmetry principles through Noether's theorem. It brilliantly unifies all the known forces in nature—except gravity—under the umbrella of a single abstract symmetry described with group theory. Gravity remains off to the side, supremely indifferent to this beautiful design. And yet Noether's theorem, the tool that allowed us to build the standard model, arose from the study of gravity, long before new exotic elementary particles such as the W and Z emerged, first from theory and then from accelerators.

The centrality of the Noether program to the standard model can be seen in the form in which the theory is almost always presented: every theoretical paper in the field uses the language of the variational calculus and of the particular branch of group theory used by Noether in her paper. The language and machinery of symmetry and conservation, in the forms exploited and developed by Noether, constitute the language and machinery of the standard model. What's more, the theories of elementary particles that together form the standard model are gauge theories, and the modern concept of the gauge theory was invented by Noether in her paper.

In his book *Physics from Symmetry*, Jakob Schwichtenberg shows how to derive all the fundamental theories of physics from symmetry. He describes Noether's theorem as "one of the most beautiful insights in the history of science" before he demonstrates how it's used in particle theories.[33]

The first half of the twentieth century brought seismic revolutions to science in the forms of the quantum and relativity theories. The rise of the standard model brought with it a different, more subtle shift in worldview. I'll explain what I mean by a slight misappropriation of some of Heisenberg's words.

Heisenberg was one of the creators of the quantum theory. He is immortalized in the Heisenberg uncertainty principle. He also thought deeply about the implications of physics for philosophy and its connection with other sciences, leaving behind an assortment of provocative articles and books. Here is something he wrote about a transformation in how to think about physics. He believed that a transformation was demanded by the emerging experimental reality and its connection with theory:

> If energy could be transformed into electron and positron pairs and vice versa, could we still ask how many particles went into the atomic nucleus? So far we had always believed in the doctrine of Democritus, which can be summarized by: "In the beginning was the particle." We had assumed that visible matter was composed of smaller units, and that, if only we divided these long enough, we should arrive at the smallest units, which Democritus had called "atoms" and which modern physicists called "elementary particles." But perhaps this entire approach had been mistaken. Perhaps there was no such thing as an indivisible particle. . . . But what was there in the beginning? A physical law, mathematics, symmetry? In the beginning was symmetry![34]

I misappropriate, because Heisenberg wrote this in 1933, and it's not about Noether's theorem—or, I would like to say now, he didn't know it was.

I think Heisenberg was being prescient. Writing before the advent of the standard model, he foresaw the central role that symmetry was to play in theory construction. His thoughts here are even related loosely to the connection between symmetry and conservation, specifically the conservation of charge. The shift in worldview that Heisenberg detected, with his sensitive philosophical antennae, is the replacement of the search for *physical objects* as the building blocks of the universe by the search for mathematical objects—specifically, symmetries.

It is Noether's theorem that makes this idea concrete, turning it from a philosophical principle into a guide that can be put into practice in the service of building and analyzing models of the universe. The theorem elucidates the structure of general relativity. It forms the basis of the standard model. Together, general relativity and the standard model explain all the known forces in the physical universe and all its fundamental constituents. Noether's theorem, therefore, is intimately intertwined with all of fundamental physics and makes much of it possible.

At an even deeper level, the method of working and thinking unlocked by the theorem, suggested by Heisenberg's musings just quoted, forever altered the flavor of theorizing in physics. In searching for extensions to the standard model (whether the extensions are empirically constrained or not), physicists almost instinctively apply the Noether program. They consider symmetry, first and last, including the hidden symmetries that may underlie the broken symmetries that we can observe, the resulting conservation laws, and the particles that those conservation laws demand. As we've seen, this reasoning flows in both directions: evidence of a new particle explained by a conservation law, implying a symmetry, which can be subsumed by a higher symmetry, leading to further conservation laws and more particles.

Hidden Symmetries

Physics teachers often reach for a pencil when trying to explain the concept of hidden or broken symmetries. There is probably an example of this archaic instrument, or something similar, on your desk. The first image that comes to mind is probably not a pencil standing upright, perfectly balanced on its point. That's because we never see such things. No matter how carefully we try to balance a pencil on its tip, it will fall over. I can be certain that any pencils on your desk right now are lying on their sides.

Now imagine a pencil balanced perfectly on its point. It exhibits no compass direction. It is perfectly indifferent to north, south, east, or west. But the actual pencils or pens on your desk are all facing in a particular direction; it can't be otherwise. The imaginary vertical pencil, balanced forever, would not be violating any laws of motion or gravity. Its symmetry in regard to horizontal direction reflects the symmetry of the laws that describe it. You will search in vain for a preferred compass direction in the downward force of gravity and Newton's laws of motion. And yet the pencils of reality all violate this symmetry, as they are all pointing somewhere. And the imaginary symmetrical pencil exists only in our theories and imaginations; we never observe it within our reality, although, theoretically, it could exist.

Physicists have adopted the term *broken symmetry* for this idea. There are hidden symmetries in nature, reflected in the symmetries of theory, but some of these symmetries are never reflected in what we see in the current state of our universe. They are broken by our imperfect reality. If we could subtract all the perturbations of real life, every breeze, vibration, or microscopic lack of smooth planar perfection in our table, and if we could place that pencil exactly vertically with infinite precision, then and only then would it remain balanced forever. If we saw such things around us, it would be easier to infer the symmetries that our theories would have to embody. What we actually observe are the effects of broken symmetries.

One of the great insights of the standard model was that the phenomenology of elementary particles reflects such broken symmetries.[35] The particle physicists realized that to apply the Noether program, they would need to imagine the hidden symmetries behind what was revealed in particle accelerators. Such hidden symmetries implied hidden conservation laws. To make empirical predictions, they needed to understand how these symmetries were broken.

Eddington

When a lazy designer needs an image of an iconic scientist, the choice is usually the same: the wrinkled face and chaotically coiffed head of

Albert Einstein. This is not the face of the man who created the ideas that turned him into that iconic scientist. That man was twenty, thirty, or fifty years younger, depending on the exact choice made by our designer. The face of the man who created the theories of relativity was unlined, handsome, and intelligent. The Einstein of the kitchen calendar and the magazine layout seems eternally, gently wise. But the younger man performed the work that eventually earned an older Einstein the Nobel Prize. The prize, however, was awarded not for relativity or for the equally decorative $E = mc^2$. It was awarded for Einstein's paper on the photoelectric effect, which helped launch the second pillar of twentieth-century physics: the quantum theory.

As described in Chapter 4, by 1918 Einstein was famous among physicists and was not without a certain notoriety in the wider world. But most people probably hadn't heard of him. But by 1919, he had become, or was rapidly on the way to becoming, one of the most famous personalities in the history of Western civilization. Journalists would hound him for his opinions on politics, religion, and sometimes even science.

What created this sudden fame and adulation? We can trace it to the persistence, vision, and courage of Arthur Eddington. Eddington was a brilliant and adventurous English astronomer with the genius to become one of the select group of scientists who could thoroughly comprehend general relativity in its early days. He understood that one of its predictions was the bending of light by gravity, and he wanted to mount an expedition to measure the apparent location of stars during an upcoming 1919 solar eclipse. As the light from the stars passed close to the sun, the light would curve, causing the stars to appear shifted from their normal positions. If the shift agreed with the predictions from Einstein's theory, it would be powerful confirmatory evidence.

After considering a multitude of factors, including the duration of the totality of the eclipse, the availability of food and water at the sites, weather patterns, transportation, and the likelihood of support from the local governments, Eddington and his collaborators settled on Sobral, in

Brazil, and the equatorial island of Príncipe, near the west coast of Africa.[36] Two observation sites doubled their chances of success; all it takes is a single cloud to ruin an expedition such as this.

Eddington had to overcome a parade of obstacles, logistic, financial, climatic, and political. Because he had been a pacifist during the war, he was already under a cloud of suspicion, and now he wanted to promote *German science*. But Eddington was not merely a brilliant and intrepid scientist; he had other talents that contributed to his ability to pull off the expedition. Among these were highly developed abilities in persuasion and explanation. He was not only a popular lecturer with a "disarming way of spinning a yarn" but a writer of textbooks, as well.[37] He did mount the two expeditions, around the time that Emmy Noether was finally allowed to habilitate at Göttingen. The results were a triumph for Einstein and for general relativity.

The front pages of the newspapers were full of excitement and giddy attempts at explanation. The world's celebration of the end of the war blended with fascination for this new kind of science. People were understandably eager to turn their eyes from the devastation around them to gaze upward. The story of the eclipse voyages was reported around the world, turning Einstein into a household name.

It was the right news at the right time. Exhausted by war, the European and American people happily took up relativity as a topic of cocktail party conversation. Einstein, bemused by it all, became a celebrity scientist of a type that had never appeared before in the world.

As a result of these events, the status of general relativity got a big boost in the physics classrooms of the world's universities. Physicists wanted to educate themselves in this new theory, and it became a standard, more or less, part of the graduate curriculum. More people took up gravitation as a topic for research.

This sudden attraction to gravitation would unavoidably have the effect of helping keep Noether's theorem alive. For the next four decades, practically the only reason for any physicist to seriously study

Noether's result and learn how to apply it to physics would have been an interest in general relativity. Few did undertake a study of Noether's work—but the critical mass of interest in Einstein's gravitational theory at least helped prevent the theorem from fading into permanent obscurity. Noether's theorem, and therefore all of modern physics, owes an enormous debt to Arthur Eddington.

8

Legacy

"If she knew how useful her mathematics had become today, she'd probably turn over in her grave."

I n 1898, the faculty of the University of Erlangen, in the city of Emmy Noether's birth and education, officially pronounced that the admission of female students to the university was a "measure that would overthrow all academic order."[1]

Since then, the rights of German women to participate in academia have steadily improved. Just ten years later, Noether received her doctorate from Erlangen. After the end of World War II, Erlangen, along with other institutions of higher education in Germany, displayed an interest in making up for lost time. They redoubled their efforts to ensure equal treatment for women and to recognize the accomplishments of those of the present and the past.

Today, Noether is celebrated in Erlangen as one of the most important citizens in its long history. In a sense, this celebration was kicked off in 1958, when the University of Erlangen memorialized the fiftieth anniversary of Noether's degree by bringing together many of her former students, and their students, to discuss the legacy and influence of her work.[2]

Noether herself, after dispensing with the mathematical problems raised in her study of general relativity in her 1918 paper, in which she proved Noether's theorem, and after being granted her habilitation largely on the strength of the theorem, lost interest in the subject. Aside from one brief, uncredited note, she never referred to her paper in any subsequent publication, nor did she suggest related topics to any of her doctoral students.[3]

This peculiar preference of Noether's stands out even in a life full of curious ironies. It's an example of how, in a fertile career replete with fundamental results, others may attach supreme value to a piece of work of little enduring interest to its creator. After the theorem, Noether returned to her consuming passions in pure mathematics, mainly algebra. For her, the theorem was the formal culmination of her efforts as a team player; she arrived at it to help her friends and colleagues— Einstein, Hilbert, and Klein—out of their confusion. The motivation arose from collegial interests, not from an internal fire. Noether's theorem, and the habilitation that it supported, punctuated her transition from an assistant in Hilbert's school to a leader in her own right: she would quickly become the center of her own school of mathematics at Göttingen and take her place in the international community of mathematicians as a forger of fresh paths.

For any typical mathematician or scientist, a result of the import of Noether's theorem would be an enviable achievement, a ticket to immortality. Yet here we are confronted with an intellect for whom this achievement was a side issue, a task of nothing more than momentary interest, completed as a favor for others—others who themselves happened to be among the giants of intellectual history. It can be a challenge to keep things in perspective, at times—or even to know what such a perspective would look like.

Nonetheless, the theorem largely defines her legacy in physics and other empirical sciences. It continues to inspire new research and find new life in a wide variety of applications that would surely have astonished its discoverer. As we will see in this chapter, Noether and her work have also had a wider cultural influence and legacy.

This book is mainly the story of an idea called Noether's theorem and the trajectory of its fortunes from its creation around 1918 up to the present day. This chapter is chiefly about how that idea continues to inspire research and find new applications. But as I've just pointed out, the theorem was of little more than passing interest to its discoverer, although it is the chief source of her enduring importance in the empirical sciences. As Noether is a giant in the world of pure mathematics, I hope I can be forgiven a slight digression in the service of briefly discussing something about her legacy in that world.

A famous mathematician of our day, Ian Stewart, who has also made fundamental contributions to theoretical physics, responded this way when asked which forgotten mathematical figure everybody ought to know about: "Emmy Noether was the greatest female mathematician of her era. She started a revolution in mathematics by focusing on abstract structure rather than computational details. This led to greater generality, and greater power. Today's mathematics—pure and applied—*couldn't exist without it.*"[4]

Here Stewart is emphasizing a point that I've alluded to repeatedly in the book. She emphasized this aspect herself, in some of her rare comments about the import of her own work. Noether saw her most enduring legacy not in a collection of results but in the dissemination of *methods of working* through the mathematical community. This stress on the approach was connected to her emphasis on abstraction and her move away from detailed computation. The new style of research arose as she turned away from the methods of her mentor Paul Gordan and her own thesis and began to enthusiastically adopt something more in the manner of David Hilbert.

As a reminder, an example of the difference between computation and abstraction can be found in the approach to showing that a particular class of equations has solutions. The Gordan school would insist that someone prove this by calculating and displaying a particular solution. Hilbert might prefer a clever proof that solutions *must exist*, without ever constructing any solutions. This was the essence of the conflict between the two men, culminating in Gordan's complaint that Hilbert's work resembled theology more than it did mathematics. Noether's thesis cranked through massive amounts of detailed calculations of this sort, work that she would later recoil from in distaste.

Mathematician Israel Kleiner, in his *History of Abstract Algebra*, lends his voice to the importance of Noether's influence in setting the direction for mathematics:

> The concepts she introduced, the results she obtained, the mode of thinking she promoted, have become part of our mathematical culture. . . .
>
> Emmy Noether was a towering figure in the evolution of abstract algebra. In fact, she was the moving spirit behind the abstract, axiomatic approach to algebra.[5]

Noether was not unique in her lack of enduring interest in her own result. Although it is the source of her fame among physicists who are aware of it, the theorem is also generally of little interest to pure mathematicians. Noether is instead a giant in this community because of the main line of her work in abstract algebra.

Sharon McGrayne puts it nicely, in her *Nobel Prize Women in Science*: Noether's theorem is "both her most famous and her least famous accomplishment."[6] Among physicists, the theorem is about the only thing she is known for (although the result remains better known than the author). Among mathematicians, however, the result is relatively obscure.

Noether's algebraic legacy includes more than an impressive host of specific results and theorems, several of which are named after her. Her influence also involves the promotion of algebra from its historical status as a calculational tool to one of the foundation stones of mathematics.[7] Noether is one of perhaps three mathematicians responsible for this transformation; today algebra is rivaled in fundamental importance only by logic itself. Noether is also largely responsible for modern algebra's abstract, "modern" nature and the methodologies surrounding work in the field.

But Noether's legacy in mathematics proper is not limited to algebra. Her impact on mathematical culture and her encouragement of a certain style of research, through her teaching and mentorship, influenced fields far from her own usual specialties.

As mathematician Peter Roquette put it, "Her methods eventually penetrated all mathematical fields, including number theory and topology"[8] and Alexandrov and Heinz Hopf's classic text *Topologie* credits Noether for creating the "algebraization" of topology.[9]

Alexandrov also gives this assessment of her stature among mathematicians during her life:

> Emmy Noether lived to see the full recognition of her ideas. . . . In 1932, at the International Congress of Mathematicians in Zurich, her accomplishments were lauded on all sides. The major survey talk that she gave at the Congress was a true triumph for the direction of research she represented. At that point she could look back upon the mathematical path she had traveled not only with an inner satisfaction, but with an awareness of her complete and unconditional recognition in the mathematical community.[10]

It's remarkable how many scholars and historians of mathematics have noticed a similar thread in Noether's influence on mathematical culture. According to Uta C. Merzbach, the Smithsonian's first curator of mathematical instruments, "The remarkable change in the language and methods of mathematics [wrought by the publications of Noether and her coauthors] is apparent upon contrasting nearly any early twentieth century mathematical publication with a contemporary one."[11]

This is quite a remarkable claim by a prominent historian of mathematics: Merzbach finds Noether's influence in nearly all of the current publication in that field.

Emmy Noether was a mathematician's mathematician, believing that mathematics should be enjoyed for its own sake, without any thought of application. In the opinion of her nephew, "If she knew how useful her mathematics had become today, she'd probably turn over in her grave."[12]

The physics world was, as we have seen, slow in catching on to the value of Noether's theorem. In the words of French mathematician Yvette Kosmann-Schwarzbach, the theorems, whose "importance remained obscure for decades, eventually acquired a considerable influence on the development of modern theoretical physics, and their history is related to numerous questions in physics, in mechanics and in mathematics."[13] She refers to "theorems": remember that what we and most others call Noether's theorem is really a set of two related theorems and their converses.

According to the views of some physicists and physics watchers, including Ruth Gregory, the head of the Department of Physics at King's College in London, Noether might have won the physics Nobel Prize had she lived longer.[14] This view is also implied in a chapter about Noether in Sharon McGrayne's *Nobel Prize Women in Science*, which deals with women who might have or should have won the prize, as well as those who did.

However, I believe that this speculation is both a distraction from the mission of elevating Noether to her rightful place in the history of science and that it is unrealistic. The prize is not awarded posthumously, and for Noether to have been a recipient, she would have had to live long enough for the importance of her contribution to be generally recognized in the physics community. The significance of her work might not have been clear until the 1960s, when the success of the standard model was generally accepted and when its dependence on the theorem was better understood. Recipients often received their prizes decades after the cited work was accomplished. Einstein, for example, received the Nobel in 1922, chiefly for one of his 1905 papers.

There is also another, perhaps more fundamental reason that the theorem would not be likely to be the basis for a Nobel Prize in physics, despite its central importance for that science. A prize in physics is unlikely because Noether's theorem is difficult to categorize as a part of physics. It's not a theory of physics, such as general relativity or quantum mechanics. To repeat a phrase I've used before, it's rather in the nature of a theory construction kit. Noether's theorem is a mathematical result that sits underneath all of physics and provides structure and interpretation. It tells physicists what theories are possible and where to look for new physics. It's so indispensable that entire edifices of physics could not exist without it. But it itself is not a physics theory. Noether's theorem is not the kind of thing that the Nobel committee is used to dealing with.

Obsessions about Nobel Prizes are best left to journalists, who take these awards more seriously than do scientists, who sometimes find having won an annoyance despite the welcome cash.[15] There is little doubt that several female physicists and astronomers were unfairly passed over for the physics prize, but Noether's name should not be among them.

The importance of Noether's theorem for physics is the main theme of this book, of course. Its legacy seems finally secure, as is the credit for its creator. Certainly, many physicists still have at best a vague idea of what the theorem is about and an even cloudier awareness of the story behind it. But as many people, including Nobel laureates and science writers who bring cutting-edge ideas to a wider audience, attest, the theorem's importance as a cornerstone of physics is enthusiastically recognized, to the benefit of Noether's place in history.

In the opinion of astrophysicist Brian Koberlein, "Emmy Noether should probably be put in the same group as Isaac Newton and Albert Einstein as one of the greatest physicists who ever lived. . . . [She] single-handedly revolutionized the way we understand physical theories."[16] Koberlein also believes that Noether was "the most brilliant mathematical physicist of all time," and her theorem's "power is hard to overstate.... Everything we study within physics depends upon Noether's theorem, from dark energy to the Higgs boson. It has transformed the way we view the cosmos, and it demonstrates the real power of mathematics when it comes to understanding the Universe."[17]

We increasingly find direct mention of Noether's theorem in research papers and textbooks, often with an aside about its immense value. This situation is a welcome contrast to the way things were near the middle of the last century, where, as described briefly in Chapter 7, an emerging field of physics depended critically on the theorem but it was rarely mentioned in any of the relevant papers.

The physicists Leon Lederman and Christopher Hill, quoted in the introduction, have more to say about the import of Noether's work for physics:

> Noether's theorem is a profound statement, perhaps running as deeply into the fabric of our psyche as the famous theorem of Pythagoras. Noether's theorem directly connects symmetry to physics, and vice versa. It frames our modern concepts about nature and *rules modern scientific methodology*. It tells us directly how symmetries govern the physical processes that define our world. For scientists, it is the guiding light to unraveling nature's mysteries, as they delve into the innermost fabric of matter, exploring the most minuscule distances of space and shortest instants of time.[18]

Symmetry has been interesting to scientists, philosophers, and artists since the beginning of recorded history. Many popularizers of science have written about how the general concept of symmetry is somehow crucial to understanding, discovery, or appreciation. But the preceding quote brings out how Noether's theorem makes the importance of symmetry precise and actionable. It explains "how symmetries govern the physical processes." After Noether's theorem, symmetry doesn't simply describe how something appears. It participates actively in the dynamics of a system, defining what behaviors are possible.

Noether's theorem is firmly embedded at the center of high-energy physics, classical mechanics, and gravitation, among other fields of physics. But physicists continue to discover new and sometimes unexpected ways to apply the theorem in areas where it has not traditionally occupied center stage. Recently, physicists studying the role of fluctuations in fluids and related systems discovered that Noether's theorem provided, as in particle physics, a guide to the construction of theories in statistical mechanics.[19] This discovery is important for physics because much of it could not exist without statistical mechanics. It is through that specialty that we deduce the large-scale properties of matter from the behavior of its constituent atoms or molecules. Aside from his theories of relativity, most of Einstein's important contributions to physics rest in large part on statistical arguments; this was the branch of physics that he turned to repeatedly to explain matter, radiation, and quantum phenomena.

As mentioned in the previous chapter, the unfinished business of modern fundamental physics, its unrealized dream, is to unify its two great models of reality—general relativity and the standard model—into one unified theory that explains everything. It's a difficult task both conceptually and mathematically, because the standard model is based on quantum mechanics, while no quanta exist in the geometry of gravitation. The two pictures of reality must be reconciled somehow, with a future theory of "quantum gravity": some way of encompassing quantum particle interactions with the non-Euclidean geometry of general relativity.

No one has yet succeeded in constructing such a theory, but recent experience suggests that Noether's theorem will remain an essential tool in the effort. The most well known attempt to create a theory of quantum gravity is string theory, along with its various offshoots. This activity remains controversial, as it's vulnerable to criticism that it has failed to produce specific predictions amenable to verification by experiment. However, it's worth noting that the mathematics of these approaches makes heavy use of Noether's theorem.

We know that despite its beauty and power, general relativity can't be the whole story of gravity. As we follow the evolution of the cosmos backward in time and approach the Big Bang, we eventually enter a realm where conditions are so extreme that they can never be reproduced in our laboratories. The physics of the early—very early—universe, the first tiny fraction of a second, will always be beyond experiment. As physics, despite the flights of imagination brought to bear on it by theorists, is at its core an empirical science, a more encompassing, unified theory, one that applied from the birth of the universe to the present, will always contain some arbitrary ingredients beyond the possibility of verification. Some things we will never know.[20]

But we do know that in the very early universe, general relativity is not sufficient. The only way to cover those highly dense conditions would be with a theory that unifies gravity with the other forces of nature, which are now described by quantum field theory. And such a theory has still eluded our most brilliant physicists.

Noether's theorem is an important guide to constructing possible alternative theories of gravity that make general relativity even more general and that may be extended to the quantum realm.[21] While groping in the dark, beyond the reach of the light of experiment, the theorem illuminates the outlines of what such theories should look like. It can do this because of its status as a theory construction kit, not tied to any particular physical hypothesis.

Particle physicists have so far not enjoyed much success in going beyond the standard model. The persistent dream is to unify all its forces into a grand unified theory (which would be grand, and unified, but which would still not encompass gravity). The program for unifying the forces of the standard model involves searching for grander symmetries. Through Noether's theorem, such symmetries would imply new conservation laws, which would need to be mediated by new particles. Some of these predicted entities are beyond the reach of current, or even foreseeable, particle accelerators—the energies required are just too big. Others are not outside our reach, and the search for these new particles can be undertaken. This was the program followed during the rise of the standard model, and it was a succession of exciting successes. But so far, the quest to enlarge the model has failed.

Physicist Sabine Hossenfelder explains the problem with contemporary particle physics in an engaging and illuminating video.[22] She points out that in their attempts to enlarge the standard model, particle physicists have made a series of predictions of new particles since the 1970s and that essentially all these predictions have failed. The basic strategy has been as described above: to find larger, more complicated symmetries that encompass the symmetries of the standard model but, in some sense, complete it and, through Noether's theorem–type arguments, to calculate the properties of the still-unobserved particles that the more elaborate symmetries require. Why has the power of Noether's theorem let them down?

Hossenfelder has a good answer, and I believe that she's correct. The history of physics presents examples of successful theories that began as more complicated elaborations of existing theories. What the successful theories all have in common is that the complications were needed to explain observations or to remedy an inconsistency in existing theory. Sometimes the theories took care of both, as with Einstein's theory of general relativity. It explained the precession of the orbit of Mercury, and it healed the inconsistencies in Newtonian gravity. General relativity explained the long-standing mystery of the identity of gravitational and inertial mass and eliminated Newton's problematic instantaneous action at a distance. The mathematics is immensely more complicated than what's needed for classical gravity, but this complication is not gratuitous. It was the natural mathematical language for the job at hand. General relativity was, and is, spectacularly successful, making a series of predictions that continue to be borne out up to the present day.

The history of the quantum theory follows the same pattern, from the early atomic theory through quantum field theory and the standard model. At each step, theories were enlarged and made more mathematically complex *when needed* to explain observations or to resolve inconsistencies. Each new model made new predictions, and these were often confirmed. The long list of triumphs bears out this pattern: Paul Dirac's prediction of antimatter, the standard model's prediction of new particles, the calculation of the fine structure constant, and other achievements were consequences of theories' grappling with observed physical reality.

The largely fruitless course followed by particle physics for the last half century has disregarded the lessons of history. Untethered from the demands of observation and experiment, brilliant theoreticians have given free range to their imaginations, creating a variety of lovely fantasies such as supersymmetry and string theory. These surmises have either produced predictions that turned out to be false, can't be tested, or produce nothing testable at all.

Instead, these theory creators are motivated by a philosophical or aesthetic desire to unify the several forces of the standard model and sometimes gravity as well. Certainly, aesthetics and metaphysics can be a guide for the scientist. Einstein's rapture about the beauty of his new theory helped convince him that it must be true. But in the absence of the constraints of observation, aesthetic concerns lead somewhere else than to successful science. Without these constraints, we have a profusion of approaches to unification, searching for a solution to a problem that doesn't exist and leading, as we've repeatedly seen, to science that doesn't work.

When we are searching for a solution to a problem that *does* exist, Noether's theorem helps show us where to look. As a theory construction tool, it can provide the scaffolding, something to stand on while you build your tower. But it has no opinions about the physics that you think might describe the world. It lies beneath the physics. To use a loose analogy from today's pervasive technology, the theorem is a key component of the operating system, which lies beneath any number of application programs. Therefore the theorem cannot tell us what the sequel to the standard model is, but it can show us where to look.

According to astrophysicist Katie Mack, "Noether's theorem is to theoretical physics what natural selection is to biology."[23] And some biologists are beginning to find insight in the theorem for their science.

I admire Mack's formulation. It perfectly captures how Noether's theorem sits underneath all of physics as an organizing principle that can inform us about what a reasonable theory might look like, no matter the details of the phenomenology under study.

One application of the theorem to biology is in the concept of *scale symmetry*. A system has scale symmetry if it looks the same at different scales. As you zoom in, the system remains unchanged.

Imagine sitting far above a planet and observing its continents with their jagged coastlines separated by seas, oceans, land bridges, canals, and so forth. Now imagine diving down for a closer look. You zero in on the edge of one of the continents, and after a while, you see a pattern that looks familiar: a set of land masses with jagged coastlines, separated by various bodies of water, land bridges, and so forth. You might think of these land masses as islands rather than continents. But at least on this planet, the pattern is eerily familiar. You've gotten closer, but everything looks the same.

Now imagine diving down to take a closer look at one of these islands and discovering the same pattern again: the island is made up of smaller islands, with bodies of water and land bridges separating them.

This example is an illustration of a type of scale symmetry. Some aspects of some systems, surprisingly, remain unchanged as you magnify your view. This kind of symmetry is different from the space-time and other symmetries we usually talk about in physics and not one of those considered by Noether. However, as her theorem deals with symmetry at an abstract level and is not restricted to symmetries of time and space (for example, the preceding examples of gauge symmetries), any newly discovered symmetry is potentially important. If it can be fit into the mathematical framework of Noether's theorem, then it can be shown to be equivalent to a conservation law. As in the case of the standard model discussed earlier, new conservation laws can lead to new predictions about nature.

It turns out that scale symmetry is more than an idea in our imaginations. Some biologists believe that some natural neural networks, including brains, exhibit some form of scale symmetry. Researchers have already connected this property with other properties, such as efficient signal transfer within the network.

The possibility of scale symmetry in the brain's neural network has now reached the level of detailed experimental investigation. Several years ago, a group of neuroscientists and physicists from London and Zürich carried out a combined experimental and theoretical study of scale invariance in animal brains.[24] Studying the question mathematically, they used Noether's theorem to derive a set of conserved quantities related to the scale invariance symmetry. These conserved quantities aren't the familiar ones, such as energy or momentum, but the scientists derived specific formulas for the new conserved quantities. They then discovered that some of the properties of animal brain activity that they measured followed the conservation laws they had derived.

This was the first study of conservation laws for scale invariance from Noether's theorem. These findings may open up avenues for research in physics and may guide future research into biological neural networks.

Of course, no one was thinking about computers in 1917. I won't try to speculate on how Noether would feel about these machines had she lived to see them. She might have found a use for them in the initial stage of her mathematical journey, which was full of tedious calculations. But after her definitive transition to more abstract forms of mathematical thought, which created in her such a distaste for her earlier work that she purportedly referred to her doctoral thesis with a mild obscenity, she probably would have had little use for computers.

What then does it say that her theorem has turned out to have a direct bearing on today's cutting-edge investigations into several areas of computer applications in science? It reminds us of the immense power of mathematical abstraction. Deep thought that uncovers unsuspected relationships between ideas has a timeless value. A theorem is true forever. We still use the Pythagorean theorem all over the world, countless times each day, millennia after it was first inscribed in the sand in ancient Greece. Noether's theorem will never become obsolete. Those who grasp it centuries from now will find new areas that it illuminates, just as today's scientists, working a century after it was proven, are applying it to fields that Noether and her friends could not possibly have foreseen.

Noether's theorem has found recent applications in quantum computing.[25] Research into quantum computers is motivated by the theoretical possibility that they could perform certain types of calculations thousands of times faster than any computer designed along familiar (classical) lines. The way to speed up computation for any type of computer is to perform as much of the work as possible in parallel, because the rate of improvement in the speed of single processors is slowing down due to the existence of ineluctable physical limits. This is why your laptop has multiple cores, and why large simulations of complex systems, such as the atmosphere, are performed on supercomputers with thousands of processors.

In a quantum computer, the parallelism is achieved by exploiting the physics of the quantum realm, where many potentialities coexist until a single one is selected by a measurement. While the computation is underway, it's essential that the computer, and its internal quantum states, remain isolated from the environment. That is, except at the beginning, when the operators define the problem, and at the end, when they read the result—during the computation, the computer must remain isolated. Interactions between these two moments can disturb or destroy the superposition of states essential for maintaining parallelism. In the context of quantum computing, these undesirable environmental interactions are referred to as *noise*. Elimination of noise is one of the central problems in the design of quantum computers.

Physicist Evan Fortunato and his coworkers have published a series of papers, beginning with Fortunato's 2002 PhD thesis, in the general area of error control in quantum computers. Fortunato's techniques are based on results derived from Noether's theorem.[26] Briefly, he has discovered that the relevant components of the noise-system interaction are characterized by certain symmetry groups. Through Noether's theorem, he can relate these symmetries with subspaces of conserved quantum states and use that relationship as a theoretical basis for control of the computation.

In his conversation with me about his work, it was clear that Fortunato considered Noether's theorem a key to the success of his research program. He suggested that others laboring without its insights sometimes had to take a long way around, while he was able to take a shortcut.

Noether's theorem has also been used to clarify problems in ordinary, nonquantum computing. Scientists and engineers routinely use computer simulations to predict the weather and climate, design airplanes, study the gravitational interactions of galaxies, and much more. In many of the algorithms underlying these simulations, the homogeneous, isotropic space in which we live is represented on some kind of grid, something that resembles the lines and boxes on graph paper.

In this computational universe, an object (an atom, an airplane, or a planet) can't glide smoothly in any direction, as it can in the real (classical) world. Instead, it's forced to jump from square to square on the predefined grid, like a piece in a game of checkers.

This grid universe doesn't have the symmetries of real space: it looks different when you rotate or translate it, just as a piece of graph paper looks different when rotated—the lines are on a slant. Therefore, the (classical) conservation laws that Noether's theorem shows us are equivalent to these spatial symmetries cannot hold within these computer simulations. Since the grid universe doesn't have, for example, spatial isotropy, it can't have conservation of angular momentum, as Noether's theorem tells us that the real (pre-Einstein) universe must have.

However, we need our simulations to be accurate. They should respect conservation of momentum and angular momentum, even if they don't fully respect the symmetries of real life. Here Noether's theorem has been useful in analyzing the effects of grids on conservation in computations and showing how to get around the absence of symmetry in certain cases.

I was once part of an undergraduate seminar whose leader, a professor of physics, confidently mentioned in passing that all of economics could be reduced to a single physics equation. Certainly this remark, which I hope was not meant to be taken entirely seriously, was just an example of that hubris of which our kind is often accused, usually by scientists in other disciplines.

Nevertheless, one approach to economics makes heavy use of the mathematics and concepts of modern physics.[27] This school of econometrics attempts to calculate invariances and conservation laws in economic systems and refers explicitly to Noether's theorem, employing it as a central tool.[28] While several "conservation laws," or invariances, such as the income-wealth conservation law, have been recognized in economics for some time, the use of Noether's theorem has revealed some of these to be special cases of more general invariances. The theorem had allowed the discovery of what the proponents of its use in economics call *hidden conservation laws*. The application of Noether's theorem in economics, and the related apparatus of group theory, variational calculus, and the machinery of analytical mechanics borrowed from physics, has led to the blossoming of a new, highly sophisticated and theoretical branch of the "dismal science." The literature of this branch of economics resembles, at a glance, treatments of the field theories of physics more than it does the familiar expositions of classical economics.

Whether these new approaches will lend economics the reliable predictive power that the discipline has long yearned for, and whether they will lead to its taking its place beside what are generally recognized as the real sciences, remains to be seen. But even if it turns out to be an extended academic exercise, the translation of the language of modern (and advanced classical) physics into the field of economic modeling is a curious and fascinating development.

Of course, things can get out of hand. One social science article that appears to be entirely serious attempts to explain something about Taiwanese politics by referring to the symmetrical behavior of a political party and its conserved objectives through application of Noether's theorem.[29]

There is clearly something infectious about the high degree of generality inherent in the way that the theorem relates symmetries with constant properties. People can and obviously will interpret symmetry and conservation in increasingly imaginative ways.

Just as quantum mechanics and relativity theory (both versions) have been subject to countless popular treatments trafficking in unforgivable vagueness and philosophical extrapolation, Noether's theorem, as it becomes increasingly better known in the general culture, will have to suffer alongside them. It's a minor annoyance compared with the gratification that would come with Noether's taking her rightful place alongside the other greats in histories of science.

Noether's struggles to participate in academic life, and the many occasions where she was thwarted or suppressed by men who were comparative mediocrities (in contrast with the conspicuously brilliant men who exerted themselves to advance her interests) recall a scene from Oscar Wilde's *An Ideal Husband*, which also happens to be a choice representation of Wilde's effortless employment of breezy irony:

LADY MARKBY. . . . *Really, this horrid House of Commons quite ruins our husbands for us. I think the Lower House by far the greatest blow to a happy married life that there has been since that terrible thing called the Higher Education of Women was invented.*

LADY CHILTERN. *Ah! it is heresy to say that in this house, Lady Markby. Robert is a great champion of the Higher Education of Women, and so, I am afraid, am I.*

MRS. CHEVELEY. *The higher education of men is what I should like to see. Men need it so sadly.*

LADY MARKBY. *They do, dear. But I am afraid such a scheme would be quite unpractical. I don't think man has much capacity for development. He has got as far as he can, and that is not far, is it? With regard to women, well, dear Gertrude, you belong to the younger generation, and I am sure it is all right if you approve of it. In my time, of course, we were taught not to understand anything. That was the old system, and wonderfully interesting it was. I assure you that the amount of things I and my poor dear sister were taught not to understand was quite extraordinary. But modern women understand everything, I am told.*

As we've seen, Noether became the center of mathematical life during the 1920s and early 1930s in Göttingen, which was then the center of mathematics in the Western world. She was one of the ten or so most important mathematicians alive in the world at the time, by any measure. Yet even in recent years, historians seem capable of a perplexing disregard not only of her importance and accomplishments but also of her very presence when they describe a milieu where she was actually dominant. I hesitate to form facile theories about the reasons for this blatant oversight, whether the causes be quirks of scholarship, psychology, or just a recurring accident, especially as this Noether-blindness is not limited to male historians. But when faced with an otherwise fine and interesting book (a collection of essays specifically about Göttingen and its importance in the development of the natural sciences, including several studies about the years during which Noether was active) and finding within its pages not a single mention of her name, one can be forgiven, I think, a certain measure of exasperation.[30]

An interesting paper in that very collection actually discusses Hilbert's battles with the Göttingen faculty over the hiring of foreigners, Jews, and, specifically, women. It names many professors who were the subjects of these skirmishes—*but omits any mention of Emmy Noether.*[31]

In addition to these oversights, several mathematicians who have studied the history of Noether's theorem have pointed out the curious fact that *special cases* of it are periodically "discovered" by modern mathematicians who believe that they have derived novel results, unaware that the problem was dispensed with, in full generality, over a century ago.

The legacy of Emmy Noether as a mathematician is obviously secure and permanent. She possesses the immortality of having her name attached to objects and results central to algebra and having indelibly influenced several other areas of mathematics. In addition, as many mathematicians have pointed out, even without these specific results, her legacy is permanently etched into the fabric of how mathematical research is carried out and discussed.

Her legacy as a role model, in light of the exemplary and courageous way she led her life in the midst of an endless cascade of cruel obstacles, will also be a permanent inspiration to all who learn her story. Her infinite generosity, impossible optimism, and utter devotion to scientific truth and the progress of mathematics define her as a unique personality in the entire human history of the exact sciences.

The moral issues that arise in Noether's treatment, first as a woman and second as a Jew at the hands of the Nazis, hardly need to be belabored. Her life is the perfect example of the potential loss that can be suffered by a civilization that withholds opportunities to flourish from segments of its population. Without Noether's work, our science would most likely not have progressed as far as it has. The tool that Noether provided future scientists with which to probe nature—from the inner workings of the atom to the structure of the cosmos—would not exist. It's exceedingly rare to encounter a figure that combines Noether's unique talents with the fortitude to push ahead in the face of a culturally entrenched disregard. Are there people, even today, with the ability to put a dent in the serious problems facing us but whose contribution we are denied because they have quite understandably decided not to pursue an exhausting existence, constantly asserting their right to be taken seriously?

My view of the history of ideas is somewhat unfashionable, as I attach what some would consider exaggerated significance to the innovations of individual minds. Lately, it's more usual to encounter the assumption that, well, if A hadn't come up with it, then B would have done so sooner or later. The idea was "in the air." This is true of some, perhaps many, of the important turning points in the twisting course of culture through time. But there also exist unique individuals with unique abilities, and there have been ideas that would simply never have existed without these people. Some of these ideas have changed everything—we would be in a different place from where we are now had it not been for these thinkers.

In no sense could we say that Noether's theorem was in the air. Without Noether, we would most likely never possess an equivalent, equally powerful and general result. Without this theorem—this theory construction kit—much of modern physics would not exist or would be in relative disarray. Our understanding of our physics heritage would be more logically uncertain; we would still labor under confused and ad hoc explanations of even basic concepts such as energy. The path forward in many areas, especially in attempts at unification, would be on even shakier ground.

I think Hilbert was right about Einstein. Others may have had a better grasp on the mathematics than he did, and predecessors such as Poincaré had many of the details at hand, and as I have pointed out, perhaps more than once, Einstein could not have reached the endpoint of general relativity without a lot of help. But without his singular and profound insights into how the universe was made, there would be no physics of the cosmos. Black holes, the Big Bang, gravitational waves, gravitational lensing, the orbit of Mercury, even the operation of GPS satellites—we would not know or understand any of it.

In Chapter 4, I told the story of Grace Chisholm Young, who received, under Felix Klein, the first PhD earned by a woman in Germany. Her son, Laurence Young, also became a mathematician. He had this to say about how the world often overlooks the importance of mathematicians to the world: "I shall say something of the significance of mathematics, and the value of mathematicians, in altering the course of history. I hope to convince you of this, although you do not find much mention of it in most history books. The great question is then, how do we produce more of these admirable beings?"[32]

So at least I have one person in my corner.

And what of the subject of this book, Noether's theorem? Clearly, any danger that this powerful result will be lost to science is well past us. It will increasingly be taught to generations of physics students at progressively earlier phases in their education. Some teachers have already told me that introducing Noether's ideas early on makes it easier to present physics as a unified body of thought and to relate it to other sciences.

We can, and probably should, emphasize symmetry arguments earlier, and with particular emphasis on how Noether's theorem shows how to relate symmetries to dynamical behavior.

I've offered a minuscule survey in this chapter to give a feel for how Noether's theorem is sending tendrils of thought and method into several disparate areas of research. The scientists who grasp its power have a chance to achieve insights and see connections that others are not equipped to see, not only within their specialties but potentially across what Hilbert would insist are the artificial chasms separating the various kingdoms of science.

After many decades of neglect, Emmy Noether is finally beginning to appear occasionally in popularizations and histories of physics, alongside Einstein and the other giants of our science. My deepest aspiration for this book is that I've nudged this process along. I trust I have convinced you that she deserves a central place in the history of science and have helped you appreciate something of the grandeur of her mathematical thought.

Above all, I hope you've joined me in absorbing the most important lesson from Emmy Noether and her life: don't forget to laugh.

Appendix

"A little lower layer"

—Captain Ahab

In this appendix, I've gathered some details whose intrusion into the main text would have made it difficult to keep the book moving forward smoothly. As I've done in the chapters, I'll cover these subjects without resorting to equations, but I will indulge in several simple diagrams. My main purpose here is to satisfy those readers with a more active curiosity about how the ideas raised by the theorem and its applications are related to each other and to the general culture of mathematics and physics. You certainly don't need much background in either of these fields of knowledge to follow this appendix. All you need is motivation, curiosity, and patience.

Abstraction

In noting the importance of Hilbert's birthplace in the historical lore of mathematics, I mentioned that the town lends its name to the

"Königsberg bridge problem."[1] The challenge is to find a path that traverses each of Königsberg's seven bridges once and only once. In other words, can you place the point of a pencil on a map of Königsberg and draw a curve that passes over each bridge exactly one time, without lifting the pencil? The arrangement of the bridges and the shape of the river that they cross make the answer anything but obvious.

Leonhard Euler, one of the (some say *the*) greatest mathematicians in history, answered this long-standing question in the negative, launching a distinct mathematical genre called *graph theory*. Euler's attack was an example of abstraction, a recurring theme in our story. In tracing the development of Noether's thought and her approach to mathematical problems, we saw how she moved from her initial training, emphasizing detailed calculations, to embrace ever more refined planes of abstraction. This transition prepared her to tackle the problems that led her to the theorem and later to become one of the most important figures in the majestic field of modern algebra, bringing to it her unique emphasis on structure and generalization.

Rather than being distracted by the particularities of rivers and bridges, Euler asks us to consider ideal structures in the abstract: networks of points and the connections between them. Today, mathematicians call such structures *graphs*. This should not be confused with the common usage of the term to mean a plot or other visualization. A graph in mathematics is a web of connections.

Euler's attack on the Königsberg bridge problem eventually grew into its own area of pure mathematical research—graph theory—which enjoys tremendous activity today. An applied offshoot of that field is known as *network science*. In this field, Königsberg's seven bridges have grown into the millions of virtual paths between internet hosts in one type of practical application or the web of connections among the victims of social media—the social network—in another.

Invariant Theory

Noether's PhD thesis adviser was nicknamed the "king of invariants," and invariant theory, the fashionable area of research at the time, is where she began her mathematical journey.

It was also the area of mathematics that had occupied Hilbert while he was at Königsberg in the 1880s.[2] He would work in this field on and off for several decades, until knocking it unconscious for several decades more by solving its principal outstanding problem.

Even a cursory understanding of what invariant theory deals with helps tremendously in our appreciation of how many of the big ideas mentioned is this book are connected with each other. The theory affords us a glimpse of the mathematical threads, sometimes nearly invisible, connecting the old theory of invariants, the turbid sea in which Noether's thesis swam, with symmetry, Einstein's two theories of relativity, Noether's theorem, modern algebra, and particle physics.

While musing on pure mathematics at Königsberg, Hilbert could not have dreamed that he was exploring a path with a treasure at its end that would have profound implications for all of science—and that this treasure would be discovered by someone else.

To better understand the context of that treasure, let's take a moment to get a feel for one of its precursors and learn a bit about what invariant theory concerned itself with. It began with an 1842 paper by George Boole, whose name is reflected in Boolean algebra, a subject that may be familiar to you if you've studied logic or computer science. Invariant theory is a deep and complicated subject that was a big part of mathematics in the second half of the nineteenth century and has continued to be elaborated upon up to today.

Despite the theory's complexity, a simple example that uses no advanced math at all can help you appreciate the concepts in play. This will be a somewhat detailed, quasi-mathematical discussion, but if you take the time to absorb it, you will be able to see how it relates to many other ideas in this book.

As the name suggests, invariant theory has something to do with things that do not change (they remain invariant) when something else happens. Figure 1 shows a pair of axes, drawn with thick black lines ending in arrowheads and labeled x and y. These are the names for the coordinates that are measured along these axes. Any location in this two-dimensional space can be specified by giving two numbers, the x- and y-coordinates. In any line graph of the budget deficit or your local temperature, a setup such as this is implicit. One location in the space is shown with a small circle; its distance from the origin, the point (0, 0) where the axes cross, is shown with a thin, dotted line. The coordinates of the circle's center are found by drawing a line from the center of the circle to the point on each axis where the line intersects at a right angle, as shown in Figure 2.

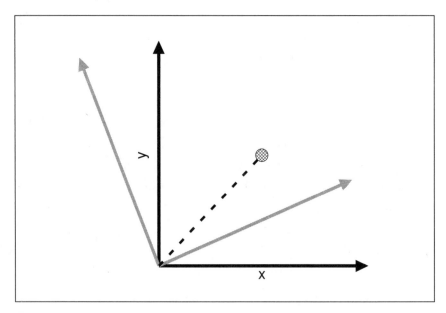

Figure 1

Now regard the other pair of axes, drawn in gray, in Figure 1. Like the darker pair, these are a normal set of coordinate axes, with a right angle between them. They are, in fact, a simple counterclockwise rotation of the original axes through some angle (about 22 degrees, but that's not important). This new set of axes defines a new coordinate system. If you want to know the coordinates of the circle in the new system, the procedure is the same: drop perpendicular lines into the new axes and see where they intersect. I suggest you do this for yourself; you'll see that the coordinates are now different. Obviously, the x- and y-coordinates of a location are not *invariants* with respect to a rotation of the coordinate system.

Take another look at the dotted line showing the distance from the circle to the origin (Figure 2). An examination of the illustration should convince you that this distance *is* an invariant: rotating the axes around the origin doesn't change the distance from the circle to the origin. Other transformations, such as shifting the axes up or down, *will* change this distance, but for now, let's just look at rotations. I specifically mention

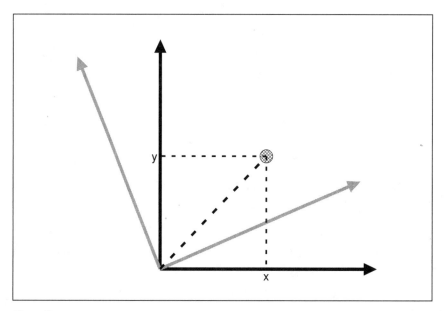

Figure 2

rotation to make something clear: we always speak of invariants with respect to some particular collection of transformations (the mathematicians out there will probably know that these collections are what they call *groups*, but this need not concern us now). The distance is, of course, not merely invariant to this particular rotation, but to *all* rotations.

There are triangles lurking within our description. Specifically, right triangles. In Figure 2, the square formed by the two coordinate axes and the two dotted perpendiculars is divided, by the line from the origin to the ball, into two right triangles. We can focus on either of these two triangles, and I will refer to them as "the triangle" from here on; either one will do. The distance shown by the dotted line connecting the origin with the ball is the imaginary triangle's hypotenuse; the distances from the origin to the ball's x- and y-coordinates, in either the original or rotated coordinate system, are the other legs of the triangle.

If you remember the Pythagorean theorem, you know that the square of the hypotenuse (the distance from the origin to the ball) is equal to the sum of the squares of the x- and y-coordinates. It's simply a fact about right triangles, an ancient fact that we all learn in high school.

The two different coordinate systems give us two different right triangles. The ones formed from the original coordinate system have legs of equal size, but the ones associated with the rotated system have unequal legs. The hypotenuses are of course the same in each, because the dotted line doesn't change; but the x- and y-coordinates are different. When we rotate the coordinate system, the x-coordinate gets larger, but the y-coordinate gets smaller. Since the Pythagorean theorem holds for the second imaginary right triangle, too, these coordinates can't change in an arbitrary way. As the x-coordinate increases while we rotate the system counterclockwise, the y-coordinate must decrease precisely so that the Pythagorean theorem continues to hold.

So far, this is all geometry. You could, if you wanted to, turn it into algebra. The Pythagorean theorem is, after all, an equation: the hypotenuse squared equals one side squared plus the other side squared. The hypotenuse is the dotted line, one side is x, and the other side is y. You can write down two equations: one before rotating the coordinate system and one after the rotation. The hypotenuse, or the left side of the equation, will be the same for both. But the coordinates, the variables on the right side, will be different. To keep things straight, we could call them x' and y'.

Written with x' and y', the second equation looks just like the first—not surprising, because both equations are just the Pythagorean theorem with different symbols. But mathematicians, physics students, and some others know how to *transform* the x and y into the x' and y'. Given a rotation of the coordinate system by a particular angle, we can write down a combination of sines and cosines of that angle and thereby write the *second* (rotated) equation with x and y instead of x' and y'. Since we know that the hypotenuse doesn't change when we rotate, we now know something new: this mess that combines the x, the y, and the sines and cosines of an angle always has the *same value*, for any angle! In other words, it is invariant. What we've really learned, rather, is that the simple expression, the sum of the square of x and the square of y, is *invariant under rotations*. We can see it from geometry, or we can prove it with algebra and trigonometry. But with the geometric insight from the Pythagorean theorem, something that was not obvious about an algebraic equation becomes something that clearly must be true.

Right about now, a mathematician would ask, Are there any *other* formulas involving x and y that are invariant under coordinate rotations? And what about other transformations? How about shifting or stretching the coordinate system? The real beginning of classical invariant theory starts with problems such as these. The activity around these questions was one of the most popular, and at times the predominant research area, but was characterized by long, laborious calculations. Perhaps "long, laborious calculations" is what most people think of when they think about mathematics, but that impression is largely a result of the oppressive way math is taught in school.[3] In reality, mathematicians search for patterns and abstractions that allow them to make progress *without* having to deal explicitly with every detail. At least, that is how mathematicians work today—but that is partly the result of David Hilbert and Emmy Noether themselves.

Hilbert got involved in invariant-theory research when his doctoral adviser, Ferdinand von Lindemann, suggested that he look into it.[4] He took Lindemann's advice and attained his doctorate in 1885 with a thesis related to the bit of math that we went into above: invariants and equations involving the squares of variables (quadratic forms).

Hilbert continued his research in invariant theory while at Königsberg. By 1888, he had wielded his characteristic self-deprecating humor to refer to himself, in a letter to his very good friend Hermann Minkowski, as "an expert invariant-theory man."[5]

In that year, Hilbert proved a theorem about invariants. The theorem put the entire field on a more abstract, sophisticated basis and made some of these laborious calculations unnecessary. It came to be called *Hilbert's basis theorem* and was one of his most famous results.

What he actually proved needn't concern us here, but it's part of an amusing anecdote in the history of mathematics. When Hilbert wrote up his proof and sent it to the prestigious *Mathematische Annalen*, it more or less automatically went to Paul Gordan himself, who happened to work for the journal and was its go-to expert on invariant theory. Gordan rejected the manuscript out of hand, recoiling from Hilbert's methods and declaring, "This is not mathematics. This is theology."[6] Perhaps worse, when von Lindemann, the professor who had supervised Hilbert's doctoral thesis, the man who had encouraged Hilbert to work on invariant theory in the first place, got hold of the paper, he called Hilbert's methods "unheimlich," which can be translated as creepy, sinister, or weird.[7]

There are fashions in mathematics and science, just as in clothing or music. Fashion leaders—and Hilbert had become one—often seem outrageous to those a bit too used to the old ways. There is also, in mathematics, an evolution of standards of proof. This development usually takes the form of a steady increase in rigor; the geometric proofs in Euclid are still considered valid, but his spare and intuitive definitions would not quite pass muster these days.

The question of validity is particularly contentious when dealing with infinities. Hilbert used the technique of proof by contradiction in his theorem. The idea is to assume the opposite of what you are trying to prove and then show that this assumption leads to a contradiction. The contradiction shows that your starting assumption must have been wrong and that, therefore, its opposite must be true, which is what you wanted to prove all along. This type of reasoning makes use of Aristotle's *law of excluded middle*: a proposition can be either true or its negation must be true—there is no other choice, and when one is true, the other must be false. Hilbert's proof by contradiction would have upset few people if he had been dealing with a finite set of objects.

The history of mathematics is full of famous proofs by contradiction. Euclid proved that there are an infinite number of prime numbers by assuming that there were only finitely many and that, therefore, you could write them all down in a list.[8] He then showed how there must be a prime that was not in the list, thereby contradicting the assumption that the list contained all the primes and proving that you can never make a finite list of them, or that there must be infinitely many.

The problem was that, in proving his basis theorem, Hilbert had wielded the proof-by-contradiction sword over infinite sets. He ended up by proving (in his opinion) the existence of something without in any way constructing that thing or showing his readers what it looked like. Up to that point in invariant theory, everything that was known to exist had been proven to exist by explicitly constructing it: hence all the laborious calculations. What Hilbert had done is what mathematicians now call a nonconstructive proof. He used logic to prove that a thing must exist, without even a hint of how to make it. Gordan's reference to theology may have been inspired by encounters with such cultural artifacts as the ontological proof of the existence of God: a trick of logic whereby God seems to poof into existence, with almost nothing said or required of His or Her nature or properties.[9]

Many mathematicians in Hilbert's time did not accept nonconstructive proofs by contradiction involving infinite sets. They insisted that the proof explicitly show that nothing was left out. This view lives on, nurtured by the school of mathematics known as *intuitionist*; the nomenclature categorizes Hilbert as a *formalist* (and the most important one, too). Hilbert, for his part, thought these objections were silly, and he went on record as having no intention of relinquishing some of his most powerful weapons, even if they made some timid mathematicians nervous.

Fortunately for him and for his paper, Gordan was overruled by Felix Klein, already a math superstar who happened to be the journal's managing editor at the time. Klein was one of those who were immediately on board with Hilbert's methods.[10] He described Hilbert's proof as "wholly simple and, therefore, logically compelling." It was this paper that created in Klein a firm resolve to get Hilbert to Göttingen as soon as possible.

Although the intuitionist–formalist controversy still has life, the end of this episode was marked by reconciliation. After a few years, Gordan found he could not deny the power and fecundity of Hilbert's methods and remarked that "even theology has its merits."[11] Hilbert, for his part, found an explicit, constructivist proof of his basis theorem—a proof that made everyone happy. Moreover, Hilbert himself would one day become the managing editor of the journal that had initially rejected his paper. Meanwhile, his basis theorem turned out to be the last interesting result in invariant theory for a couple of decades. After its most important open question had been solved, few people wanted to try to extract a living from its soil. It took someone of the caliber of Hilbert's student Hermann Weyl to revive the field, and not until 1939.[12] Invariant theory lives on, but it never looked the same after Hilbert had rearranged the furniture.

Emmy Noether's PhD thesis sat squarely in the invariant-theory tradition of explicit construction. A tour de force of voluminous, detailed calculations, the thesis contained a table of over three hundred explicitly calculated invariant formulas. It was a thesis that her adviser, Paul Gordan (and her father, Max, who also happened to be a significant laborer in these same fields) could be happy with, as it was very much in his style of research. It was soon published in a math journal, taking up sixty-seven pages.[13]

A few years after completing her thesis and while still in Erlangen, Noether discovered Hilbert's results. This discovery began her transformation as she was won over to his methods.[14] Noether began to break away from the Gordan style; she adopted a more abstract approach and eventually forged methods all her own.

Special Relativity

In Chapter 1, I mentioned how Hermann Minkowski discovered how to describe the transformations of space and time used in Einstein's special theory of relativity as a rotation in a four-dimensional space-time. The transformation itself, called the Lorentz transformation, amounts to a separate formula for the three spatial coordinates and another for time: hence the four dimensions. Minkowski's mathematically profound insight was to see that these separate formulas, when viewed together, are expressions of a single geometrical concept: a rotation.

(The obscure fact that Voigt was the first to discover the Lorentz transformation was mentioned in Chapter 2.[15] It was also known to Poincaré, and, of course, Lorentz. However, they don't get credit for special relativity, which is far more than a transformation formula.)

The ideas are the same if we limit ourselves to one space dimension plus time. This simpler approach covers all the examples of trains and airport walkways and allows us to visualize what's happening with simple pictures. And we've already seen this example in Figures 1 and 2. Look again at these diagrams of a rotating coordinate system. There we had two space coordinates, which we called x and y. Now replace y with time, but not exactly: you have to multiply the time coordinate by c, the velocity of light, and by i, the imaginary unit. It becomes an imaginary time axis, scaled by the velocity of light. The other axis is still just x. Minkowski showed that the transformation that holds between reference frames, where time slows down and space contracts, is completely described by a rotation of this space-time coordinate system. As the train speeds up, you rotate by a larger angle. Points on the train, when measured from the platform, get new space-time coordinates that we can calculate from the rotation.

Just as the length from the origin to the ball is invariant under spatial rotations, the "length" in space-time, that is, the distance between two "events" in its coordinate system, is invariant. Thus the ideas of invariant theory make contact with the new physics of relativity. Minkowski had discovered a hidden symmetry in the equations of special relativity, a symmetry that Einstein hadn't seen.

When we rotate the coordinate system in our original example, I mentioned that we could write the transformed x and y in terms of the original coordinates, using some sines and cosines. The transformation mixes up the coordinates, in the sense that the new x, x', has some of the old y mixed into it, and y' has some of the old x. It's the nature of rotations. That means that when Minkowski's space-time coordinate system is rotated, the new spatial coordinate will have some time mixed into it, and the transformed time will contain an admixture of space.

The invariant, the length of a line, illustrated in the preceding diagrams is of course also an invariant in Einstein's theory. This invariant would turn out to be important in our understanding of the physics of relativity.

Appendix

Hilbert and Euclid's Geometry

As described in Chapter 2, Hilbert developed a lifelong interest in the formal structure of mathematics. One of the first fruits of his obsession was his reformulation of Euclid's geometry.[16] Particularly interesting for us is Hilbert's penetration into the meaning of Euclid's fifth axiom, the parallel postulate.

But first, as an example of the lucidity of his treatment of axioms and definitions, let's see what Hilbert does with the first axiom in Euclid's work. In the original, the Greek geometer simply says that you can draw a straight line from any point to any other point.[17] Hilbert judged the first axiom to be imprecise. He made it clear that the line that Euclid had in mind was *unique*: any two points define one and only one line that passes between them.[18] The critical quality of uniqueness was crucial for the axiomatic structure of the theory.

More interesting, especially as we'll see later, is how he replaced Euclid's rather labyrinthine fifth postulate with this:

> In a plane α there can be drawn through any point A, lying outside of a straight line a, one and only one straight line which does not intersect the line a. This straight line is called the parallel to a through the given point A.[19]

I've illustrated the postulate in Figure 3 to make it clear what Hilbert means. (Similar restatements of the fifth postulate were put forth by John Playfair in 1795 and as early as the fifth century CE by Proclus.[20]) Hilbert distilled Euclid's version into a simpler and clearer statement that also introduces the concept of *parallelism*. Interestingly, this axiom has also recently been called by such names as *Euclid's parallel postulate*, although the original version doesn't mention parallelism—simply implies it. But the attentive reader of Euclid will notice that in his definition 23, he defines parallel lines and uses similar wording in his fifth axiom.[21]

The possibility of constructing alternative geometries in which the parallel postulate is replaced with something else is crucial for the mathematics underlying general relativity. The curvature of space-time that Einstein introduced into physics means that Euclid's ancient geometry does not describe the universe in which we live. Hilbert's work in putting the axiomatic structure of that geometrical system in order clarified for mathematicians which axioms might be replaced while maintaining consistency.

Einstein, Gravitation, and Geometry

In Chapter 2, we saw how Einstein, while struggling to find a mathematical language to express his new ideas about the space-time geometry

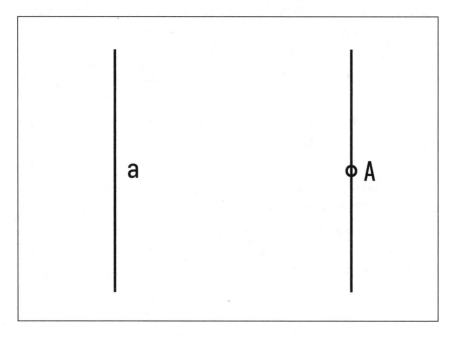

Figure 3

of reality, was rescued by his friend Marcel Grossmann, who discovered the needed advanced geometry in the library. Here I'll briefly identify that geometry more specifically.

It was Grossmann who realized that some form of Riemannian geometry would serve for general relativity. More specifically, he discovered, in the library stacks, more recent elaborations of Riemann's geometry by Tullio Levi-Civita and other mathematicians.[22]

Bernhard Riemann himself had enjoyed a career at Göttingen.[23] Rather astoundingly, the mathematician, who died in 1866, had anticipated in his comments on the geometry of the universe some of Einstein's thought.[24]

During Einstein's final push to apply the mathematics of Levi-Civita to make his equations work, after his 1915 visit to Göttingen, he was no longer collaborating with Grossmann. His tutor became Emmy Noether, who worked out a great deal of the tensor math to make general relativity into a consistent and viable theory.[25]

To gain a schematic idea of what these new forms of geometry entailed, let's return to Euclid's fifth postulate, preferably in the previously quoted, superior Hilbert version.[26] The postulate says that there is one and only one line, a unique line, parallel to an existing line, that passes through the point exterior to the first line. It seems intuitively, obviously true (perhaps after a moment of thought). Euclidean geometry is defined by this axiom plus the other four—and, of course, by the definitions. The geometry that Einstein needed was born when some creative mathematicians asked themselves what would happen if they replaced the fifth postulate with something else. Here there are basically two choices: there can be *no* line passing through the point A that is parallel to the first line, or there can be *more than one* (which, for technical reasons, means that there must be an infinite number of such lines).

These ideas are not as bizarre as they may seem at first. Most of our intuitions about geometry are formed by thinking about geometry on flat surfaces: so-called plane geometry. There, the postulate seems obviously true. However, geometry on a curved surface, such as on the surface of the earth, is different.

Before we consider geometry on a curved surface, we have to be clear about what a "line" is going to be. Since lines in general are not straight on a curved surface, and since, on some types of surfaces, they can't extend to infinity, the infinite straight lines of Euclid's geometry need to be replaced with something more appropriate. On a plane, we know that a straight line is the shortest path between two points. The natural extension of the line concept to a curved surface is to retain the idea of the shortest path. We call these paths *geodesics*. Where Euclid talks about a line, we'll substitute a geodesic on the general surface. That's how the mathematicians who developed *non-Euclidean geometry*, the collection of geometries that replaced the parallel postulate with an alternative, defined the analogue to the straight line.

You may be familiar with the concept of the *great circle* on the globe. The great circle is the path, approximately, that commercial airplanes take between two points, because it's the shortest distance that follows the surface of the earth. In other words, the great circle is the geodesic on a sphere. And on the sphere, Euclid's parallel postulate doesn't work. One of the alternatives considered by our exploratory mathematicians does, though: on the surface of the globe, there are *no* geodesics (lines) parallel to a given geodesic.[27] Any two great circles intersect. Geometry on a sphere is just one example. It turned out that non-Euclidean geometry, at first considered little more than a logical curiosity, was the geometry that described life on curved surfaces.

Einstein had used the geometrical language developed in the nineteenth century to describe gravity as an alteration in the metrical structure of four-dimensional space-time caused by the presence of mass (or energy). Two masses moved toward each other not because of an instantaneous force acting at a distance but as a result of their following the shortest path, or geodesic, through space-time. The apparent force of gravity was analogous to the fictional forces that arise from a change of reference frame, such as the centrifugal "force." This idea automatically resolved the mystery of the two types of masses: Why was the gravitational mass exactly the same as the inertial mass? Just as the fictional centrifugal force you feel when going around a bend has nothing to do with your gravitational mass, the geometry-based "forces" in general relativity explained the action of gravity completely. There was only one mass, and the mystery of the identity of gravitational and inertial masses evaporated.

Local Energy Conservation

We saw how Noether discovered her theorem when she studied the problem of energy conservation in general relativity. More specifically, she was trying to resolve paradoxes that Hilbert thought he had found when examining the status of local energy conservation in the theory. Here we'll take a more detailed look at the local conservation laws described briefly in Chapter 3.

Suppose we draw a circle around a region of space, as in Figure 4. The circle might contain electric and magnetic fields, water in motion, or something else. If we keep track of the energy (E_f) flowing out of (and into) the perimeter of the circle during a certain interval of time, then that total transfer of energy must be equal to the amount that the total energy inside the circle (E_v) has changed. This is the essence of local energy conservation and just means that energy is not created or destroyed but is just moved around.

It's really a simple idea. To relate it to the physics of smooth fields and substances, such as electric fields or water, we allow these circles to become infinitely small, and fill all of space with them. Then the local energy condition holds at every point in space, and it becomes a differential equation. Differential equations are essentially the language in which we write all of our physics ideas. They relate infinitesimal changes in space and time to other infinitesimal changes. In this case, the differential equation relates an infinitesimal change in the energy in an infinitely small volume to its flow into or out of that volume. It's a perfectly strict accounting off all the energy, and its movements, in systems where that energy can be spread over space.

In Maxwell's theory of electricity and magnetism, the electric and magnetic fields themselves—not just charged masses—can contain energy (and momentum, and even angular momentum, as mentioned in Chapter 3). We need a real differential equation, not just a counting up of discrete objects, to describe this.

This type of differential equation, the *continuity equation*, appears all over the place in physics. It is used whenever we need to say that something behaves as a substance that can flow from point to point and that may (or may not!) be allowed to "build up" or to be evacuated from any

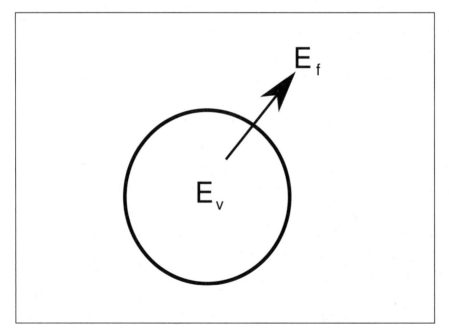

Figure 4

particular area but that is never created or destroyed: the total amount of the substance never changes. When applied to energy, the continuity equation expresses the local conservation of energy. It also appears in hydrodynamics applied to mass, to the actual substance that flows, where it expresses the more ancient idea of conservation of mass. Here we have two versions of the equation, depending on the type of fluid that we are considering. In the compressible version, mass, like energy, is allowed to accumulate and the density can therefore increase at any point. In the incompressible version, the patterns of flow are restricted so that the density remains the same everywhere. Local energy conservation is described by a differential equation that tells us that energy flows around like a compressible fluid: at any particular location, energy can decrease or be depleted, but this change must be accounted for by its movement across nearby boundaries. Like air, energy can flow from place to place but can't appear out of nowhere.

It was this problem, when applied to general relativity, that perplexed Hilbert. He was unable to show that the field equations for gravitation were consistent with local energy conservation, that is, with a proper continuity equation for energy in the four-dimensional space-time that Einstein had invented for the universe to live in.

Explaining how Noether proved her theorem in any detail is far beyond the scope of this book. However, for readers who are interested in following an argument that lends plausibility to a class of special cases of the theorem, Richard Feynman has bent his strong explanatory talents to the task, with some success. He has produced a comparatively simple, partly pictorial argument that, while not actually proving anything, at least demonstrates why a simplified, special case of Noether's theorem might be true.[28] (Feynman mentions that he's treating a special case of a general relationship between symmetries and conservation laws, but it's not clear that he's aware of the existence of a general theorem, and he never mentions Noether.) I recommend this intriguing "picture proof" to readers interested in following a semimathematical argument.

Hilbert and Quantum Mechanics

One anecdote makes astonishingly clear how Hilbert's powerful mathematical intuition could have had a profound impact on the history of twentieth-century physics, had the physicists around him been ready for it. Although the episode occurred in a slightly later period and digresses into quantum mechanics, the slight detour is worthwhile.

Heisenberg had more or less stumbled across the first mature theory of quantum mechanics when he worked out the rules for how quantum states evolve and how they behave when subject to measurement. These rules perplexed him, as they seemed to involve an odd mathematical machinery. When he showed the theory to Max Born (whose name now appears, along with Heisenberg's, in every quantum mechanics textbook), Born recognized the patterns and explained to Heisenberg that this was the arithmetic of matrices. This description made things clearer, from a calculational point of view, but the two men were still confused about *what* these matrices had to do with the atomic physics problems they were trying to model. So they went to see the master.

They showed their work to Hilbert, who immediately told them that if they wanted to understand things better, they could look for the differential equation for which these matrices were characteristic solutions (*eigenfunctions*).[29] Heisenberg and Born didn't understand why Hilbert was going on about differential equations. They concluded that they were wasting their time with him, and never followed up with his suggestion. If they had, they would have discovered the famous Schrödinger equation before Erwin Schrödinger found it. This differential equation is the second formulation of quantum mechanics, called wave mechanics;

Heisenberg's version is called matrix mechanics. Even after the publication of wave mechanics, it took some time before the two theories were proved equivalent. But Hilbert had penetrated to the heart of the matter almost at a glance and doubtless could have derived Schrödinger's equation as a diverting exercise. But he probably thought that if it were important to the physicists, they would work it out themselves. Perhaps he can be forgiven his crack about physics being "too hard for physicists."

And here is something fascinating about Schrödinger's equation and Noether's theorem, something that neither Noether nor Schrödinger would have had any idea about. It turns out that Schrödinger's equation, which lies at the center of the wave picture of quantum mechanics, contains a certain simple mathematical symmetry. It's similar to the gauge symmetry in the equations of electromagnetism described in Chapter 3. There I mentioned that Noether's theorem shows how the gauge symmetry in Maxwell's equations is equivalent to the conservation of charge. In Schrödinger's equation, Noether's theorem shows that its symmetry is equivalent to the conservation of probability.[30]

Modern Algebra

Chapter 6 dealt in part with the subject of Noether's teaching and research supervision during her tragically brief time at Bryn Mawr. Her main interest was in promulgating and advancing abstract algebra.

Here I can provide a more concrete picture of what this important area of mathematics deals with, using an example. In keeping with the theme of much of this book, the example has something to do with symmetry.

Think about a square. Feel free to draw it, or use your imagination. Now think about rotating the square about its center. Rotation by some random angle will make the picture of the square look different, but rotation by certain special angles will leave it looking the same. These are the angles related to the symmetry of the square. It's probably obvious that a rotation by zero degrees leaves the square unchanged and that a rotation by a full turn, or 360 degrees, also has no effect. Those two rotations are true of any shape. (Now you need to decide whether a rotation means rotating clockwise or counterclockwise, but it doesn't matter, as long as you're consistent.) In addition, rotation by 90 degrees leads to no change, and that's because of the square's symmetry. It certainly won't be true of a picture of a house. Other angles that leave the square looking the same are the multiples of 90 degrees: 180 and 270 degrees (Figure 5).

Different shapes would lead to different special angles. If we were considering a regular hexagon rather than a square, the angles that would leave it unaltered would include 30 degrees.

So far, we've identified five angles that leave the square unchanged: 0, 90, 180, 270, and 360 degrees. But there are four more: the negatives of the nonzero angles. These negative-degree rotations mean rotating counterclockwise if you decided that positive angles meant clockwise rotations, or *vice versa*. That gives us nine angles defined by the requirement that they leave the square unchanged.

So far this may seem to have more to do with geometry than anything you might call "algebra." But the next step is to think of these nine rotations as objects in themselves, and the act of applying a rotation as

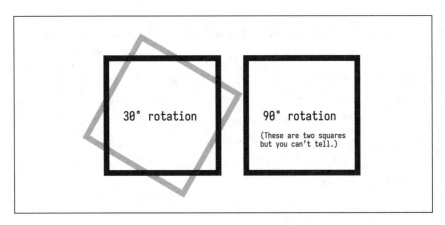

Figure 5

an operation. You can operate with one of the objects on another one: applying a 90-degree rotation to a 180-degree rotation is the same as a 270-degree rotation, for example. Applying a 0-degree rotation to anything leaves it the same. Also, for every rotation, there is another one that undoes it, leading to the 0-degree rotation; it's just the negative of the one we started with. For example, if we operate with the −90-degree rotation on the 90-degree rotation, that's the same as a 0-degree rotation. One last thing: it makes no difference how we associate the operations. In other words, if we combine a 90-degree turn with a 180-degree turn and take the result, which is a 270-degree turn, and apply it to another 90-degree turn, we get a final result of 360 degrees. But we could also combine the last two rotations first and operate on that result with the first rotation, and we would still get the same result. This indifference to how operations are combined is called the *associative property*. This property may seem obvious, but it doesn't hold for everything! (Another property, that the *order* of operations doesn't matter, happens to be true here, but it's *not* true in, for example, rotations in three dimensions.)

These nine objects, along with the associative property, plus the fact that we have a zero element that doesn't do anything, *plus* the fact that every object has a "reverse" object that gets us the zero element, plus one more thing, define a particular kind of structure. That one more thing is the fact that our little universe of rotations is *closed*: any of the nine rotations can act on any other, and we get another one of the nine. We never "escape" from the system.

This special structure is called a *group*. It's the first structure studied in a course of abstract algebra.

The point of abstract algebra is abstraction. The example of rotations of a square is just an example, to make the ideas concrete. The group itself, the structure of the nine elements, is the real object of study. When you prove things about this group, you now know something about any other group with the same structure, or about all groups, depending on the assumptions behind your theorem. Another example of a group is the set of all whole numbers with the operation of addition. That's an infinite group, while our square rotation group is a finite group. But there's another kind of addition, where we "wrap around" a finite set of numbers (it's called *modular addition*). If we limit our set to the nine elements consisting of zero and the integers one to four in the positive and negative directions, and wrap around, so adding one to four gets us back to zero, we have a finite group with the same structure as the group of rotations of the square. Any general statement that we prove to be true about this group is true for either example of it or for any other example with the same structure. That is what is meant by *abstraction*: algebra abstracts structures from particular instances and proves general propositions about those structures.

Gravity Revisited

In Chapter 2, we saw the importance that Einstein attached to his new theory's ability to correctly crank out the perihelion drift of Mercury. Newton's theory of gravity, which had ruled the universe for centuries, predicted eternally closed, stable elliptical orbits for the planets, aside from the known perturbations caused by other planets in their vicinities. The drift of the axis of Mercury's orbital ellipse, standing in insolent contradiction to the classical law of gravity, had been a mystery since Newton's time. Like the other irregularities in the motions of the solar system, it had been reasonable to look for the cause in the perturbative influence of an undiscovered planet, but one had never been found.[31] The puzzle wasn't resolved until general relativity—until Einstein took the bold step of claiming that it was not our catalog of planets that was incomplete, but our understanding of gravity itself.

Besides the crucial connections we saw between Einstein's theory of gravity and Noether's theorem, there is also a fascinating connection between Newtonian gravity, planetary orbits, and Noether's theorem. This relationship illuminates how the theorem can complete our understanding of an area of physics by exposing a hidden symmetry.[32]

Let's pretend that the universe contains only the sun and Mercury (or any other planet). Remember that the planet will orbit the sun, tracing out a path in the form of an ellipse. As we're now experts about symmetries and conservation laws, we know that this sun-Mercury system will conserve energy, momentum, and angular momentum. These conservation laws must be true from Noether's theorem, because of time translation, spatial translation, and orientation symmetries—symmetries buried deep inside our intuitions about time and space and reflected in the classical physics equations that describe the solar system. (We also know that none of these things are really true, because they were replaced by the intuition-busting theory of general relativity. But we're talking about the old physics now.)

We can write a simple formula involving the angular momentum of the system and the separation between the two bodies (the sun and

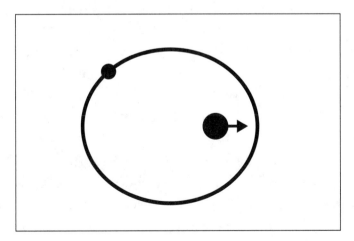

Figure 6

Mercury, in our example). This formula will define a vector, which is like an arrow with a specific length and direction, because the separation is a vector pointing between the two bodies, and so is the angular momentum (Figure 6).

This vector is known as the Runge-Lenz vector (or sometimes the Laplace-Runge-Lenz vector, and sometimes just the Lenz vector). It points from the sun toward the planet's perihelion, and its length is proportional to the eccentricity (the ovalness) of the orbit's ellipse.

Curiously enough, it happens to be conserved, as proved in 1799 by Pierre-Simon Laplace.[33] As the planet orbits its star, the vector doesn't change, which means that both its length and its direction remain fixed forever, even while the separation between the two bodies is constantly changing. Even though Carl Runge and Emil Lenz came along much later, they somehow got promoted to share the credit. And worse, Laplace's name is usually left off entirely. So it goes, far too often, with the naming of things in science and mathematics.

For the classical force of gravity, we have this extra conservation law in addition to energy, momentum, and angular momentum. We know that Noether's theorem tells us that every conservation law is equivalent to some symmetry. For the three familiar conservation laws that hold for planetary orbits, their associated symmetries are also familiar: energy conservation is equivalent to symmetry in time, and so on. This new conservation law must also have an associated symmetry, and the theorem makes it possible to work out exactly what this is. It's not an obvious space-time symmetry, as we have for the other three, but an abstract symmetry described by a particular Lie group. In fact, it's the symmetry of rotations in four-dimensional space. (More strictly speaking, since the conserved object is a vector, it gives us three conserved quantities, its components in three-dimensional space. The angular momentum likewise gives us three conserved quantities. When we combine them, the six conserved numbers together lead, through Noether's theorem, to the four-dimensional rotation symmetry.)

The theorem has told us something new about a very old problem. For centuries, it contained a hidden symmetry. In fact, Kepler's ellipses are projections, in the three spatial dimensions in which we live, of a particle traveling on a hypersphere in four dimensions.

And what good is it to know this, interesting as it may be?

For many mathematicians and physicists, and perhaps for some normal people as well, "interesting" may be good enough! But there is far more. The hidden symmetry uncovered by Noether's theorem, of four-dimensional rotations, is the same symmetry that we use to analyze various systems in quantum mechanics, in particular its canonical problem of the hydrogen atom.

Early in the history of our knowledge of atoms, we discovered that they were arranged with a heavy, positively charged central part, the nucleus, surrounded by much lighter negative particles, the electrons, with a lot of empty space separating the two components. The electrons had to be orbiting the nucleus, or else they would just fall in—just as the planets need to orbit their sun. So the atom was imagined to be like a tiny solar system.

However, this view led to a paradox. The electrons are charged, and everyone (every physicist, anyway) knew that when you accelerate a charged particle, it radiates electromagnetic energy. Acceleration refers to a change in velocity, which means a change in speed or direction. An orbiting thing is constantly changing at least the direction in which it's traveling, so an orbiting electron is being constantly accelerated. (The way we create radio signals is to accelerate electrons back and forth inside an antenna.) This state of constant acceleration meant that the atom could not survive for long. As the electrons continually emitted energy, the energy of their orbits would have to decrease, because of conservation, just as the orbiting neutron stars that we met in Chapter 3 will eventually crash into each other because of the energy lost through gravitational radiation. Soon the electrons would crash into the nucleus. As this didn't happen, the first quantum physicists made up some ad hoc rules to save the atom from annihilation. They said that the electron orbits were "quantized": the tiny particles could only exist in a discrete list of energy levels. The electrons were simply not allowed to continuously emit energy and spiral into the nucleus. Problem solved!

Since hydrogen is the simplest atom, with one proton and one electron, it received the most study, and its allowed list of energy levels was known in detail. If you heat up a bit of hydrogen gas, it will glow, just as hot things do in general. But it won't glow with a rainbow of light, with a continuous spectrum. If you send the light through an instrument that acts like a prism, separating light into its components, you will see individual colors, with gaps between them—or, rather, individual wavelengths, as the radiation coming out of the hot hydrogen is not limited to visible light. This set of discrete wavelengths is called the *spectrum* of the hydrogen atom. The wavelengths correspond to the energy levels of the electrons in the atom; quantum theory relates the wavelength of a photon to its energy.

As quantum mechanics developed, the quaint model of miniature solar systems was replaced by ideas such as the *wave function*. Electrons were no longer tiny planets with well-defined paths through space. Instead, they and all the other minuscule inhabitants of the quantum realm escaped our classical intuitions of what physical things should be like. The wave function gave us the *probability* of observing them in any particular place, if we chanced to look. When we weren't looking, they were . . . everywhere and nowhere. The wave-function approach solved the radiation paradox in a far more sophisticated way. The electrons weren't traveling along orbits. They didn't have any paths in the classical sense at all.

Although the electrons were not following orbital paths, they did still inhabit a quantized set of allowed energy levels. They had to, because the observed discrete spectrum was still there! Much of modern quantum theory makes heavy use of group theory, the mathematics of symmetry. One way physicists can calculate the energy levels of the hydrogen atom is to combine the Schrödinger probability waves with the Lie group of rotations in four dimensions. Although the geometry is four-dimensional, the *group* is six-dimensional, because it takes six parameters to define the rotation. *Why* there are six parameters is too technical to explain here, but it has to do with the size of the matrices that represent the rotation operators. In two dimensions, when we're rotating a circle, the Lie group is one-dimensional; in three dimensions, when we're rotating a sphere, the group is also three-dimensional; and in higher spatial dimensions, the dimensionality of the group increases rapidly. Although the mathematical reasons for the six dimensions of the group are a little obscure, the physical reasons, given above, are less so; they're related to the six numbers needed to define two conserved vectors in the planetary system.

Quantum mechanics allowed the calculation of hydrogen's energy levels by applying this Lie group structure. The result agreed, and continues to agree, spectacularly with experiment.

Here, then, is one final gift that Noether's theorem offers us when we apply it to the problem of a planet orbiting its star: the theorem reveals a hidden, four-dimensional symmetry that connects the classical problem with the quantum physics of the hydrogen atom and brings us, by way of an abstract mathematical detour, full circle. For this atom began, in the imaginations of scientists, as an orbiting system itself; when that model was recognized as untenable, it was replaced with mysterious clouds of probability, subject to the laws of the quantum theory and the symmetry group of rotations in four-dimensional space that yield its spectrum. Now the theorem shows us that Kepler's orbits are themselves manifestations of this same rotation group.

This is what theoretical physicists live for: the sudden realization of deep connections between apparently vastly disparate phenomena, unified through profound and beautiful mathematical structures. Noether's result is a powerful tool for seeing inside our various models of the universe and illuminating their hidden connections.

One more connection to modern physics is worth noting. Since the Runge-Lenz vector is conserved, and it points to the perihelion, then the position of the perihelion can never change. The planet must make perfect, closed ellipses forever. This is a property of the gravitational force, as well as the spring force; a force with even a minutely different dependence on distance will not create closed orbits. But remember that the perihelion of Mercury *does* change. Calculating its precession was a challenge for Einstein and his new theory of gravity and was a historic triumph when he succeeded. The perihelion of Mercury precesses because in general relativity, the gravitational force is not exactly the force of Newtonian gravity, which gives rise to perfect ellipses. The precession is measurable because the closest planet to the sun orbits within a stronger gravitational field, hence a more strongly curved space-time, than the other planets. Noether's discovery showed how energy conservation in general relativity needed to be replaced with conservation of the canonical energy-momentum tensor, as we saw in Chapter 3, because the symmetries in Einstein's universe are different from the symmetries in Newton's universe. The story of the Runge-Lenz vector adds another facet: the alteration of Newton's gravitational force law destroys the four-dimensional rotational symmetry that Noether's theorem uncovers, but the breaking of this symmetry breaks the conservation of the Runge-Lenz vector. And this allows the perihelion to precess.

By the way, the Runge of the Runge-Lenz vector is the same mathematical physicist who, as mentioned in Chapter 3, thought he had found a way out of the energy conservation problem in general relativity, but whose idea was shot down by Emmy Noether.

Representation Theory

Chapter 7 went into some detail about how the authors of the standard model depended on Noether's theorem to create modern particle physics. As this book is devoted to the influence of that theorem, I didn't discuss another part of mathematics essential to quantum field theory in general and the standard model in particular: representation theory.

In Chapter 3, I touched on group theory, as it's an important component of Noether's theorem. I also mentioned Noether's legacy in modern algebra, which starts with groups and goes on to study progressively more complex structures. And earlier in this appendix, I related the curious incident where Hilbert tried to tell Heisenberg and Born how to discover the Schrödinger equation, which the two physicists promptly ignored. All these observations and events are related. The objects that Heisenberg had stumbled on and which Born had recognized as matrices are the representation of a group that encapsulates the symmetry of the simple quantum system that Heisenberg was analyzing. Every group (of a certain kind that happens to be important in quantum mechanics) can be represented by a set of matrices, with a special multiplication rule familiar from linear algebra. The representation of groups by matrices became its own mathematical subject, called representation theory.

Weyl had something to say about this in his eulogy for Noether: "This is the aim of the theory of representations. The theory of noncommutative algebras and their representations was built up by Emmy Noether in a new unified, purely conceptual manner."[34]

The noncommutative property that Weyl mentions is the characteristic of a mathematical operator where the result depends on the order of its operands. For example, the addition or multiplication of numbers is commutative, because $2 + 3$ is the same as $3 + 2$, and 2×3 is equal to 3×2. But subtraction and division are clearly noncommutative. The certain kind of group that I mentioned earlier is one where its multiplication operator is noncommutative. This is the case with the conventional multiplication of matrices, so representation theory in general is a theory of noncommutative structures.

Noether put representation theory on a modern footing, as she had done for much of the rest of abstract algebra, and this area of mathematics became one of the foundation stones of the standard model. While this contribution may not be as profound as Noether's theorem, mainly because representation theory had developed to a certain point without Noether's help, it's another example of her fingerprints on parts of physics that didn't exist until long after her death.

Extensions of Noether's Theorem

The theorem itself is the subject of continuing investigation. The idea is to widen its arena of applicability beyond Noether's original context of continuous Lie groups. Some of this work includes efforts to subsume noncontinuous symmetries, such as reflections, transitions between discrete states, or grids with reduced spatial symmetries. Another effort is related to the interest in using the techniques of Lagrangian mechanics in systems that dissipate energy (usually into heat). The embrace of dissipation is important because huge areas of physics and engineering relate to such things as the flow of real fluids, where friction (viscosity) causes mechanical energy to be lost in the form of heat. The energy is lost because it can't be recovered and converted back into the energy of motion. Because of the one-way direction of the transformation of energy, the Lagrangian formulation of mechanics doesn't apply in the usual way, so that, tragically for us fluid dynamicists, Noether's theorem can't be applied directly to our problems. However, if there were some way to extend the usual idea of a Lagrangian to dissipative systems, we would be able to exploit the theorem.[35]

Noether's Legacy in Pure Mathematics

Mathematicians and scientists celebrate Noether's life and legacy regularly, in forms ranging from commemorative T-shirts to lectures to an array of programs and scholarships named after her. To give some flavor of the range of these activities, I present here a limited, random selection.

The American Mathematical Society has a yearly Noether Lecture: "The lecture honors Emmy Noether (1882–1935), one of the great mathematicians of her time. She worked and struggled for what she loved and believed in. Her life and work remain a tremendous inspiration."[36]

The International Mathematical Union also has a Noether Lecture that "honors women who have made fundamental and sustained contributions to the mathematical sciences."[37]

In 2017, mathematics professor Anthony Bonato established the Emmy Noether Scholarship in the mathematics program at Ryerson University (now called Toronto Metropolitan University). In his announcement, he describes Noether's influence: "Every mathematics undergraduate learns of Noetherian rings and Noetherian modules. . . . She's one of the most important mathematicians in modern times, yet her name is unknown to many non-mathematicians. Along with giants like Curie, Turing, and Einstein, she should be one of our modern scientific heroes."[38]

In 2013, the European Physical Society established the EPS Emmy Noether Distinction for Women in Physics to "bring noteworthy women physicists to the wider attention of the scientific community, policy makers and the general public" and to "identify role models that will help to attract women to a career in physics."[39]

The Institute of International Education's Scholar Rescue Fund inaugurated the Emmy Noether Chair in 2020 to support "outstanding women scholars who face threats to their lives and careers": "the Chair honors the prominent mathematician Dr. Emmy Noether, who was assisted by IIE's Emergency Committee in Aid of Displaced Foreign Scholars after fleeing Nazi Germany in 1933. It supports annually, and in perpetuity, an outstanding woman scholar who, like Noether, overcame significant barriers to pursue her scholarly work."[40]

The Perimeter Institute, an independent Canadian research institute for theoretical physics and outreach, has a set of "Emmy Noether initiatives." They include the Emmy Noether Circle, the Emmy Noether Council, and an emerging talent fund, a fellowship, and scholarship awards all named after Noether.[41]

The Amalie Emmy Noether Fellowship in Applied Mathematics and Scientific Computing at Brookhaven National Laboratory supports PhD scientists for two years of research.[42]

This list continues with a few more entries from other various spheres of life and academics:

A recent paper on gauge theory mentions that "Noether's 1918 theorem relating infinitesimal 'global' symmetries to conservation laws, is a cherished cornerstone of modern theoretical physics."[43]

Quantum field theory and the standard model have Noether's theorem at the core. As Steven Weinberg, one of the architects of the standard model, explains, "But why the Lagrangian formalism? Why do we enumerate possible theories by giving their Lagrangians . . . ? I think the reason for this is that it is only in the Lagrangian formalism (or more generally the action formalism) that symmetries imply the existence of Lie algebras of suitable quantum operators, and you need these Lie algebras to make sensible quantum theories. . . . If you start with a Lorentz invariant Lagrangian density then because of Noether's theorem the Lorentz invariance of the S-matrix is automatic."[44]

Thomas Garrity's recent text on Maxwell's equations also explains important connections: "Noether's theorem is one of the most important results of the twentieth century. It provides a link among mechanics . . . , symmetries . . . , and conservation laws."[45]

And from *Quantum Field Theory in a Nutshell*, by Anthony Zee: "We now come to one of the most profound observations in theoretical physics, namely, Noether's theorem, which states that a conserved current is associated with each generator of a continuous symmetry."

The interesting book *Chaotic Harmony: A Dialog About Physics, Complexity and Life* contains a chapter about symmetry and Noether's theorem, touching on isospin (an aspect of quantum mechanics important in the development of the standard model) and the complex subject of angular momentum conservation in general relativity.[46]

In 1960, the city of Erlangen named a street in a new residential district the Noetherstrasse.[47]

In the opinion of the novelist Ransom Stephens, Noether's theorem is "the most important discovery of humanity" and "perhaps the most important discovery in the encyclopedia of human understanding."[48] Before he became a novelist, Stephens was a particle physicist, which certainly helps explain the awe with which he regards the theorem. He understands that his discipline could not exist without it. Stephens has written an entertaining novel in which a character named Emmy Nutter is loosely inspired by Noether; the book contains something of an explication of Noether's theorem.[49]

A short-lived Myspace page called Noether Symmetry had a set of rather repetitive electronic music songs. It's remembered with affection here.[50]

There is a crater on the moon named Nöther (the way her surname is sometimes spelled) and an asteroid named 7001 Noether.

The Hotel EMC2, in Chicago, commissioned mathematician and artist Eugenia Cheng to turn its lobby into an installation in honor of Noether and her mathematics.[51]

Finally, the internet comic *XKCD*, beloved by nerds everywhere, mentions Noether now and then.[52]

These signposts in the culture are encouraging. They suggest that Noether is slowly emerging from a figure obscure to those outside the halls of math and physics and beginning to enjoy a small degree of popular renown. Or at least that people who, like Randall Munroe, the author of *XKCD* and several works of popular science, have one foot in the technical world and the other comfortably planted in the popular culture know that if they mention her name, their audiences will smile.

Quantum Computing

Chapter 8 described in general terms how Noether's theorem has found an application in quantum computing. For those desiring a bit more detail, we first need to know more about the ideas behind this new type of computer.

In all classical (normal, nonquantum) digital computers, the fundamental unit of computation is the *bit*, a state that is either on or off. We think of the bit as a number, a 1 or a 0, or a logical value, a true or a false. I qualified the description of classical computers with the term *digital* because there do exist such things as *analog computers*, which represent a physical system by modeling it with continuous quantities, rather than digital on-or-off values. Analog computers are not as common these days as they once were, having been almost entirely replaced by digital machines because of their convenience and flexibility. One example that you might have seen is the common slide rule, now supplanted by its digital descendant, the electronic calculator.

In the modern digital computer, the bit is represented by an internal voltage, with, typically, a high voltage (often near five volts) standing in for 1, and a low voltage, near zero, meaning 0. In some computers, bits are not represented electrically at all. Computers can be designed with light or the flow of liquids, for example. When I was a child, I had an educational toy in the form of a functioning, programmable digital computer made almost entirely of plastic, with some small metal bars and springs. You pushed the computation along manually by manipulating a handle in and out to cycle the mechanism.

The point is that the familiar electronic computer is just one possible representation of the classical digital architecture. Any substrate that stores and manipulates distinct 1 and 0 states, bits, can be a digital computer. Whether they're built with integrated circuits or sand, they're all based on the same concept.

A quantum computer is something different. Instead of the bit, quantum computers represent state with something that's been dubbed a *qubit*. A qubit is binary, like a classical bit, in that it has two possible states, but they are quantum states. In physics, a quantum state is something such as the spin direction of an electron or the polarization of a photon. And one reason quantum behavior is so different from our familiar reality is that quantum states interact and behave differently from classical states. Qubits don't behave like bits.

You've probably heard of Schrödinger's cat. The poor imaginary creature, kept in a box in perfect isolation from the environment, has been, with some probability, exposed to a fatal poison by a mechanism also completely within its isolated world. Classical thinking would lead us naturally to say that the cat is either alive or dead, with the specified probability. But that is wrong. The cat exists in the *superposition* of two states, neither alive nor dead, until we open the box and observe its condition. This act of *measurement* causes the combined state to collapse or *decohere* into one or the other possibility. This is the essence of the standard account of quantum reality, the *Copenhagen interpretation*. It is *not* the case that the cat was *either* alive or dead, and we just didn't know until we looked. Rather, the cat was both and neither, and its state of life or death *did not exist* until the interaction with the larger environment that we call a quantum measurement.

This state of affairs is the source of perennial dyspepsia among both physicists and others who have tried to come to terms with the quantum universe, with the consequence that the Copenhagen interpretation is continually challenged by alternatives.[53] So far, none have supplanted it as the workaday set of concepts used by scientists in planning and interpreting experiments. In Chapter 8 I described how Noether's theorem can help control noise in quantum computers. Now we can be more concrete: noise causes the computer's internal superposition of states to decohere.

Grids and Simulations

The classical way to use a computer to simulate, say, the flow of water is to solve the equations for fluid dynamics on a computational *grid*. The grid divides up space (in one, two, or three dimensions) into cells. Imagine drawing a picture on a piece of graph paper by coloring in the boxes rather than drawing it freehand by sketching your curves at will on a blank page. The common technique for using a computer to solve physics equations is to use a virtual piece of graph paper.

This way, by using grids, we can replace the mathematics of infinitesimal changes over infinitesimal distances by small but finite changes over a set of discrete locations. This lets us approximate calculus by arithmetic (computers can't do calculus, leaving aside "computer algebra," but they're very good at arithmetic). Grids are not the only way to approach such equations on a computer, but they're still one of the main methods.

The introduction of grids creates some of its own complications, however. For one thing, it replaces the continuous symmetries of reality—spatial translation symmetry and the perfect isotropy of space—with a construct, an artificial universe, that has reduced symmetries.

For example, you can rotate your blank piece of paper any way you want and (ignoring its edges) begin your drawing; the rotation makes no

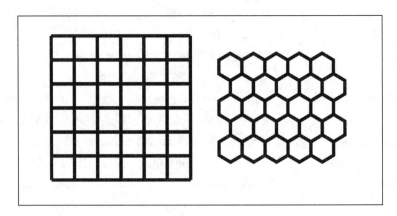

Figure 7

difference. But that's not true with graph paper: the little squares will be in a different orientation. The grid makes a difference.

Figure 7 shows two sections of two different computational grids. These represent tiny portions of what would be a grid typically thousands of times as large. In a simulation, we define the variables of our physical fields at the center of each cell (or perhaps at the boundaries, or a combination of both, but in any of these cases, the symmetry issues are the same).

The grid on the left, a square grid, has a reduced symmetry compared with physical reality. The space defined by this grid is not isotropic but only has 90-degree rotational symmetry. The image on the right is a hexagonal grid. These are not used as commonly in simulations as square grids, but they're seen sometimes. It has a higher symmetry than the square grid, in the sense that its 60-degree rotational symmetry is a bit smoother, a bit closer to a circle. Put another way, if you rotate this grid, it comes back to its original configuration six times in a complete circuit, whereas the square grid only lies on top of itself four times.

Since isotropy, as we've learned, is equivalent to angular momentum according to Noether's theorem, we might expect simulations using these grids to have some problems conserving angular momentum when the systems they are meant to represent should obey this conservation law. In fact, *grid artifacts* are a persistent problem in grid-based numerical simulations. There are ways to reduce the effects and account for these problems, but they usually can't be entirely eliminated. Noether's theorem is a useful tool in analyzing and predicting the effects of broken computational symmetries on the resulting faithfulness to conservation laws.

Conway's Game of Life is a computer program designed to illustrate how complex behavior can emerge from a simple set of rules.[54] It's a fascinating demonstration of something that can strike the observer as an arena of interacting organisms living on a computer grid. The program began as a recreation but was later put to use as an experimental computational tool, becoming the basis for new simulation techniques.

In Conway's Game of Life, the arena is a square grid like the one on the left of Figure 7. Each cell in the grid can be alive or dead, with life usually indicated by coloring the cell black and dead by leaving it blank. The game is initialized by bringing some of the cells to life, and then it proceeds on its own, evolving according to some simple rules. The rules specify under what conditions a cell will change state between life and death: cells can die, spring to life, or remain unchanged. The transitions are determined only by the number of live and dead neighbors for each cell. For example, one rule is that if a dead cell has exactly three live neighbors, it will come to life.

These simple rules lead to surprisingly complex and varied behavior. At first, the program had no practical purpose except as a provocative demonstration of emerging complexity and unexpected results. But it turned out to be such a good illustration of just these things that it's still studied intensively by computer scientists, mathematicians, and physicists.

Figure 8 shows eight frames from a Game of Life simulation that I programmed on my laptop.

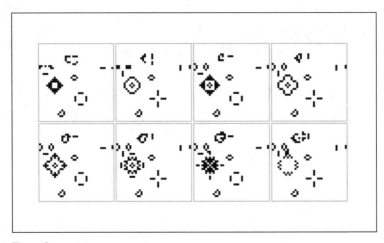

Figure 8

The simple definition for this system makes it simple to program, as well; my version takes less than fifty lines of Julia code to both do the calculation and draw the pictures (and there is a famous version in a single line of APL).[55]

The suggestive variety of behavior that can emerge from this simple set of rules has mesmerized some researchers into believing that they have discovered a new kind of science. It also inspired more down-to-earth investigations of the possibility of using this type of system, with a binary (alive or dead) state defined on a grid and following a short list of rules, as the basis of an efficient technique for simulating continua, especially fluids. The computational approach was inspired by an analogy with nature: we believe that macroscopic behavior, such as the flow of water, arises from the collective action of a large number of particles interacting according to some set of rules. Although the rules may require some sophisticated math to describe and be hard to find, they are, in some sense, simple and probably finite. So why not use a computer to calculate the interactions of a large number of computer particles and see if the average behavior looks like moving water?

Some early attempts to implement this idea were promising. The class of computational systems, resembling in a general way Conway's Game of Life and used for physical (or other) simulations, is known as *cellular automata* or *lattice gases*. However, we know now that many of these types of simple computational systems can create images that visually resemble natural flows but are not correct. (The efficiency of the calculations, however, can make them useful in computer animation for games or movies, where appearance is more important than scientific accuracy.)

The reason for the lack of accuracy is the fact that the square grid lacks the symmetry properties of real space. Since the square grid is not isotropic, Noether's theorem reminds us that the simulation won't conserve angular momentum. This lack of conservation leads to results that, when averaged to extract the large-scale behavior, fail to obey the

Navier-Stokes equations—that is, the fluid equations that should govern that behavior.

A breakthrough was reached in 1986 by three scientists, two working in France and one in the United States.[56] They created a correct approximation of the Navier-Stokes equations on a hexagonal grid like the one on the right side of Figure 7. Although the hexagonal grid may be a step closer to true isotropy than the square grid, it's obvious that it still lacks the symmetry of physical space. The surprising 1986 result was that hexagonal grids, in concert with certain averaging procedures, could be rigorously shown to have symmetry properties that led to a correct solution of the physical equations. The same demonstration showed why square symmetry could not work.

These authors used the hexagonal symmetry group and related it to terms in the equations connected to the conservation laws, using mathematical ideas drawn from Noether's work. This is one example of the extension of Noether's theorem from its original context involving continuous symmetries to systems with discrete symmetries.

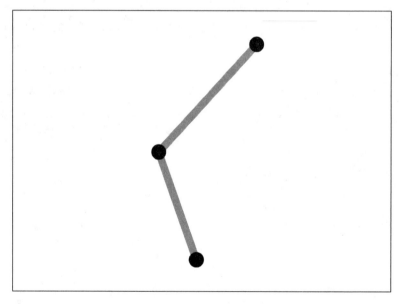

Figure 9

Robotics and Control Theory

Imagine you have a simple device, such as the one shown Figure 9, made from two rigid bars connected at a pivot (the middle black circle). The device might be two-dimensional, in which case the pivot is a type of hinge, or be free to move around in three dimensions, making the pivot a ball joint.

In either case, the configuration, or state, of the device can be completely described by a set of numbers (two or three) for each of the black circles, giving the coordinates of its position. That would make nine numbers in the three-dimensional case. But do we need so many numbers? Not quite. Think about the simpler two-dimensional case first. If we first fix the bottom circle in space with its two spatial coordinates (typically x and y), then the middle circle's position can be completely defined by *one* number: the angle of the bottom bar. And then the top circle needs one additional angle to specify its location. In three dimensions, we have two angles for each bar. The number of coordinates has been reduced by the *constraints* on the system: the balls are constrained by how they can move.

Constraints simplify a mechanical system in this way: they reduce the number of coordinates that we need to describe its configuration. But they also complicate the analysis of a system in mechanics, because it's often quite involved to calculate the forces that the constraints exert on the moving parts of the system to keep them . . . constrained. Since these forces are hard to calculate, the dynamics of the system is also hard to figure out.

Beginning about thirty years ago, and continuing to pick up steam until the present moment, research has been underway into the control of robots based on Noether's theorem. The robots, and related devices, are modeled as dynamical systems with constraints. Figure 9 illustrates a simple system, yet it's still complicated to analyze in detail. Imagine having to control the position of the top circle by only manipulating the bottom circle, perhaps having to balance the whole thing against gravity. It would be a difficult juggling trick.

Noether's theorem simplifies the analysis of these systems by exploiting the additional symmetries created by the constraints and the permitted control forces.[57] From these symmetries, the theorem is used to derive the conservation laws special to the device under study; knowing these conservation laws permits the calculation of an optimal control strategy. This approach potentially allows control of the robot with fewer calculations and can greatly ease analysis, making possible some designs that would seem intractable if they required a detailed force calculation. This general approach has been applied to a large range of related problems, including robotic locomotion and the orientation of satellites.

Pedagogy

To ensure Noether's place in the history of science commensurate with her accomplishments or at least to remove the veil of invisibility from her reputation, we must change how physics is taught at the beginning and intermediate levels. Unfortunately, little seems to have improved since I was an undergraduate. Today's physics student can easily survive to a bachelor's degree without knowing anything about Noether's theorem or having heard of its discoverer.

To be sure, in advanced works on classical mechanics it has never been unusual to encounter a treatment or mention of special cases of Noether's theorem, if not always by name. For example, a section of V. I. Arnold's advanced but influential book, *Mathematical Methods of Classical Mechanics*, contains a section on Noether's theorem.

However, when I refer to a shift in teaching, I have in mind something a bit more substantial than the sprinkling of historical anecdote as a seasoning over the conventional treatment of introductory physics. Symmetry principles, and their dynamical implications through Noether's theorem, can be introduced early in the curriculum. This early introduction can not only provide a more unified approach to the subject but also better prepare the student for more advanced studies in gravitation and particle physics.

A handful of articles have appeared explaining aspects of Noether's theorem at the advanced undergraduate level, attempting to remedy its relative neglect in standard textbooks. Some of these take innovative approaches, including the use of interactive software to make the connection between symmetries and conservation laws more intuitive for the student.[58]

Acknowledgments

Thank you to my children, to whom this volume is proudly dedicated. May you achieve everything you dream of and even those things that neither you nor I can yet imagine.

I am grateful to the editors of *Ars Technica* for publishing an article of mine about Emmy Noether. The article led to the existence of this book, thanks to Evan Fortunato, who noticed it and brought it to the attention of my agent, Susan Rabiner. Thanks also to Dr. Fortunato for sharing with me details about the use of Noether's theorem in his research.

Thanks to the indefatigable Ms. Rabiner for inviting me to expand the story into a book and working with me on its plan and approach.

I deeply appreciate the assistance of the librarians in charge of Bryn Mawr's special collections in helping me to navigate several boxes of archival material related to Emmy Noether's tenure at their beautiful campus. The government employees staffing the Library of Congress in Washington, DC, do not receive enough acknowledgment for their patient dedication to the noble profession of the librarian. I am happy to do so now and to thank them for their help in locating both routine and obscure materials.

Another dedicated group whose work goes largely unacknowledged comprises the developers of free software, which I use exclusively. Warm thanks to the creators and maintainers of Pandoc, gnuplot, Julia and its Luxor drawing package, and Linux and the civilization of Gnu tools, too numerous to list. These are the instruments that I depended on to make the diagrams in this book, to organize references, to write, and to conduct research.

My indispensable and precious friend Karina Mejía was my cheerleader throughout this project. Her empathy, curiosity, and encouragement played a big part in supplying the fuel that propelled this book to completion.

I owe a real debt to the sincere interest shown in my work by my fellow scientist and author Kevin Jensen. I walk away from our too-infrequent meetings with regret but, without fail, harboring new ideas and buoyed by a new impetus. I am proud to be your friend.

Thank you to Cinthya Lara for patiently taking dozens of photographs in an attempt to make me sufficiently presentable for the book jacket. Your rejuvenating companionship during the rare intervals when I was not writing helped me to confront each new stretch of the road ahead with equanimity.

Thank you to my sisters Melicia and Meredith and their families for their interest and hospitality at various times during this book's gestation.

For the joy of convivial comradeship, valuable leavening of these past months of labors, I am grateful to Mileisha Zelaya, Juán Calderon, and Patricia Munguía, and a special warm thanks to my tolerant teacher and faithful friend Mónica Toro.

Finally, I am eternally grateful to Emmy Noether for leading a life that cannot fail to be an inspiration for all who learn something about it, for all time.

Notes

Introduction

1. Frank Wilczek, *A Beautiful Question: Finding Nature's Deep Design* (New York: Penguin Press, 2015), 280.

2. Leon M. Lederman and Christopher T. Hill, *Symmetry and the Beautiful Universe* (Amherst, NY: Prometheus Books, 2004).

3. Brian Greene (@bgreene), "Emmy Noether's theorem is . . . ," tweet on X (formerly Twitter), March 23, 2017, https://twitter.com/bgreene/status/844768 785248641027.

Chapter 1: Intersecting Paths

1. Auguste Dick, *Emmy Noether: 1882–1935* (Boston: Birkhäuser, 1981), 9.

2. Eliza Bisbee Duffey, *The Ladies' and Gentlemen's Etiquette* (Philadelphia: Porter and Coates, 1877).

3. Cliff Stoll, "Acme Klein Bottle," www.kleinbottle.com.

4. Renate Tobies, "The Development of Göttingen into the Prussian Centre of Mathematics and the Exact Sciences," in *Göttingen and the Development of the Natural Sciences*, ed. Nicolaas Rupke (Göttingen: Wallstein, 2002), 116–142.

5. Max Born, Hedwig Born, and Albert Einstein, *The Born–Einstein Letters: Correspondence Between Albert Einstein and Max and Hedwig Born from 1916–1955, with Commentaries by Max Born*, trans. Irene Born (New York: Macmillan, 1971), 13.

6. Constance Reid, *Hilbert* (Berlin; Heidelberg: Springer Verlag, 1970), 43.

7. Ibid., 46.

8. Ingrid Daubechies and Shannon Hughes, "Konigsberg Bridge Problem," in Math Alive: Graph Theory, Princeton University, last updated August 2008, http://web.math.princeton.edu/math_alive/5/Lab1/Konigsberg.html.

9. David Hilbert, *The Foundations of Geometry*, trans. E. J. Townsend (LaSalle, IL: Open Court, 1902).

10. David W. Lewis, "David Hilbert and the Theory of Algebraic Invariants," *Irish Mathematical Society Bulletin* 33 (1994): 42–54, www.maths.tcd.ie/pub/ims/bull33/bull33_42-54.pdf.

11. Reid, *Hilbert*, 14.

12. Ibid., 46.

13. Dick, *Emmy Noether: 1882–1935*, 11.

14. Leon M. Lederman and Christopher T. Hill, *Symmetry and the Beautiful Universe* (Amherst, NY: Prometheus Books, 2004), 69.

15. Dick, *Emmy Noether: 1882–1935*, 13.

16. Ibid., 122.

17. Reid, *Hilbert*, 48.

18. Ibid., 49.

19. Ibid., 52.

20. Ibid., 53.

21. Hermann Weyl, "David Hilbert and His Mathematical Work," *Bulletin of the American Mathematical Society* 50 (1944).

22. Reid, *Hilbert*, 33.

23. Ibid., 94.

24. Paul Hoffman, *The Man Who Loved Only Numbers: The Story of Paul Erdos and the Search for Mathematical Truth* (New York: Hyperion, 1998), 95.

25. Reid, *Hilbert*, 53.

26. Ibid., 109.

27. Laurence Young, *Mathematicians and Their Times: History of Mathematics and Mathematics of History* (Amsterdam; New York: North-Holland, 2012), 238.

28. Reid, *Hilbert*, 91.

29. Ibid., 69.

30. Dick, *Emmy Noether: 1882–1935*, 13.

31. Ibid., 14.

32. Ibid., 120.

33. Abraham Pais, *Subtle Is the Lord: The Science and the Life of Albert Einstein* (Oxford, UK: Oxford University Press, 1982), 46.

34. Dick, *Emmy Noether: 1882–1935*, 112.

35. Ibid., 152.

36. John Gribbin, *Einstein's Masterwork: 1915 and the General Theory of Relativity* (New York: Pegasus Books, 2017), 117.

37. Michio Kaku, *Einstein's Cosmos: How Albert Einstein's Vision Transformed Our Understanding of Space and Time* (New York: W. W. Norton, 2004), 74.

38. Albert Einstein et al., *The Collected Papers of Albert Einstein*, vol. 8, *The Berlin Years: Correspondence, 1914–1918* (English translation supplement), trans. Ann M. Hentschel (Princeton, NJ: Princeton University Press, 1998), 172, http://einstein papers.press.princeton.edu/vol8-trans/.

39. Gribbin, *Einstein's Masterwork*, 117.

40. Einstein et al., *Collected Papers*, 236.

41. Dick, *Emmy Noether: 1882–1935*, 19.

42. Ibid., 23.

43. Reid, *Hilbert*, 87.

44. Ibid., 102.

45. Ibid., 104.

46. Ibid., 108.

47. STANDS4 Network, "What Does Geheimrat Mean?," Definitions & Translations, STANDS4 Network, n.d., www.definitions.net/definition/Geheimrat.

48. Reid, *Hilbert*, 88.

49. Norbert Wiener, *I Am a Mathematician* (Cambridge, MA: MIT Press, 1956), 96.

50. Reid, *Hilbert*, 68.

51. Ibid., 119.

52. Ibid., 85.

53. Ibid., 90.

54. Ibid., 100.

55. Lederman and Hill, *Symmetry and the Beautiful Universe*, 70.

56. David E. Rowe and Robert Schulmann, "General Relativity in the Context of Weimar Culture," preprint 456, Max-Planck-Inst. für Wissenschaftsgeschichte, Berlin, 2014, www.mpiwg-berlin.mpg.de/Preprints/P456.PDF.

57. Kaku, *Einstein's Cosmos*, 108.

58. Sharon Bertsch McGrayne, *Nobel Prize Women in Science: Their Lives, Struggles, and Momentous Discoveries* (Washington, DC: Joseph Henry Press, 2001).

59. Peter Freund, *A Passion for Discovery* (Hackensack, NJ: World Scientific, 2007).

60. Hilbert, *Foundations of Geometry*.

61. Leo Corry, "Hilbert and Physics (1900–1915)," in *The Symbolic Universe: Geometry and Physics 1890-1930*, ed. Jeremy J. Gray (Oxford, UK; New York: Oxford University Press, 1999), 145.

62. Uta C. Merzbach, "Emmy Noether: Historical Contexts," in *Emmy Noether in Bryn Mawr: Proceedings of a Symposium*, ed. Bhama Srinivasan and Judith Sally (New York: Springer, 2011), 161–171.

63. Thomas Levenson, *The Hunt for Vulcan: . . . and How Albert Einstein Destroyed a Planet, Discovered Relativity, and Deciphered the Universe* (New York: Random House, 2015).

64. "Which Falls Faster—a Feather or a Hammer?," *BBC Teach*, 2024, www.bbc.co.uk/teach/terrific-scientific/KS2/zd9r2sg.

65. E. M. Lifshitz and L. D. Landau, *The Classical Theory of Fields*, 4th ed. (Woburn, MA: Butterworth-Heinemann, 1980).

66. The Economist, "The Most Beautiful Theory," *Economist*, November 28, 2015, www.economist.com/science-and-technology/2015/11/28/the-most-beautiful-theory.

67. James Overduin, "Einstein's Spacetime," Stanford University, November 2007, https://einstein.stanford.edu/SPACETIME/spacetime2.html.

68. Matthew Stanley, *Einstein's War: How Relativity Triumphed amid the Vicious Nationalism of World War I* (New York: Dutton, 2019), 49.

69. Albert Einstein et al., *The Collected Papers of Albert Einstein*, vol. 5, *The Swiss Years: Correspondence, 1902–1914* (Princeton, NJ: Princeton University Press, 1993), 188, https://einsteinpapers.press.princeton.edu/vol5-doc/238. Einstein's original words were "Nun bin ich also auch ein offizieller von der Gilde der Huren etc."

70. Dick, *Emmy Noether: 1882–1935*, 24.

Chapter 2: Gravity

1. Lee Phillips, "General Relativity: 100 Years of the Most Beautiful Theory Ever Created," *Ars Technica*, December 3, 2015, http://arstechnica.com/science/2015/12/general-relativity-100-years-of-the-most-beautiful-theory-ever-created/.

2. Constance Reid, *Hilbert* (Berlin; Heidelberg: Springer Verlag, 1970), 103.

3. "Leo Corry's Website," www.leocorry.com.

4. Leo Corry, "Hilbert and Physics (1900–1915)," in *The Symbolic Universe: Geometry and Physics 1890–1930*, ed. Jeremy J. Gray (Oxford, UK: Oxford University Press, 1999), 145–188.

5. Peder Olesen Larsen and Markus von Ins, "The Rate of Growth in Scientific Publication and the Decline in Coverage Provided by Science Citation Index," *Scientometrics* 83, no. 3 (2010): 575–603, doi:10.1007/s11192-010-0202-z.

6. Corry, "Hilbert and Physics (1900–1915)."

7. David Hilbert, *The Foundations of Geometry*, trans. E. J. Townsend (LaSalle, IL: Open Court, 1902).

8. Leo Corry, *David Hilbert and the Axiomatization of Physics (1898–1918)* (New York: Springer, 2010), 379.

9. Euclid, *Euclid's Elements of Geometry*, trans. Richard Fitzpatrick (Richard Fitzpatrick, 2007), www.cs.umb.edu/~eb/370/euclid/EuclidBook1.pdf.

10. Albert Einstein et al., *The Collected Papers of Albert Einstein*, vol. 8, *The Berlin Years: Correspondence, 1914–1918* (English translation supplement), trans. Ann M. Hentschel (Princeton, NJ: Princeton University Press, 1998), 265, http://einsteinpapers.press.princeton.edu/vol8-trans/.

11. Ibid., 418.

12. Abraham Pais, *Subtle Is the Lord: The Science and the Life of Albert Einstein* (Oxford, UK: Oxford University Press, 1982), 151.

13. Ibid., 152.

14. Ibid., 216.

15. Ibid., 209.

16. Ibid., 213.

17. David E. Rowe, "Einstein Meets Hilbert: At the Crossroads of Physics and Mathematics," *Physics in Perspective* 3 (2001): 379–424, https://link.springer.com /article/10.1007/PL00000538.

18. Simon Singh, "The Wolfskehl Prize: The History Behind the Most Famous Prize in Mathematics," Open University Maths Society, October 1997, https://simon singh.net/media/articles/maths-and-science/the-wolfskehl-prize/.

19. Pais, *Subtle Is the Lord*, 236.

20. Rowe, "Einstein Meets Hilbert."

21. Einstein et al., *Collected Papers*, 122.

22. David E. Rowe, "Making Mathematics in an Oral Culture: Göttingen in the Era of Klein and Hilbert," *Science in Context* 17, nos. 1–2 (2004): 118, http://journals .cambridge.org/action/displayAbstract?fromPage=online&aid=229047&file -Id=S0269889704000067.

23. Vasudevan Mukunth, "Beyond the Surface of Einstein's Relativity Lay a Chimerical Geometry," *Wire*, September 10, 2015, https://thewire.in/science/beyond -the-surface-of-einsteins-relativity-lay-a-chimerical-geometry.

24. Jeremy John Gray, "Bernhard Riemann," *Encyclopedia Britannica*, 2022, www.britannica.com/biography/Bernhard-Riemann.

25. Hermann Weyl, *Die Idee Der Riemannschen Fläche* (Leipzig and Berlin: B. G. Teubner, 1913).

26. Albert Einstein, "Maxwell's Influence on the Development of the Conception of Physical Reality," in *James Clerk Maxwell: A Commemoration Volume 1831–1931*, by J. J. Thomson et al. (Cambridge, UK; New York: Cambridge University Press, 1931), 66–73, https://books.google.hn/books?id=zzYOP1EXyroC&dq=einstein ++maxwell&lr=&source=gbs_navlinks_s.

27. The Economist, "The Most Beautiful Theory," *Economist*, November 28, 2015, www.economist.com/science-and-technology/2015/11/28/the-most-beautiful -theory.

28. Michio Kaku, *Einstein's Cosmos: How Albert Einstein's Vision Transformed Our Understanding of Space and Time* (New York: W. W. Norton, 2004), 108.

29. Corry, "Hilbert and Physics (1900–1915)."

30. Einstein et al., *Collected Papers*, 568.

31. Rowe, "Einstein Meets Hilbert."

32. Tessa E. S. Charlesworth and Mahzarin R. Banaji, "Gender in Science, Technology, Engineering, and Mathematics: Issues, Causes, Solutions," *Journal of Neuroscience* 39, no. 37 (2019): 7228–7243, www.ncbi.nlm.nih.gov/pmc/articles /PMC6759027/.

33. Auguste Dick, *Emmy Noether: 1882–1935* (Boston: Birkhäuser, 1981), 31.

34. Sharon Bertsch McGrayne, *Nobel Prize Women in Science: Their Lives, Struggles, and Momentous Discoveries* (Washington, DC: Joseph Henry Press, 2001), 71.

35. Einstein et al., *Collected Papers*, 109.

36. Ibid., 111.

37. Ibid., 116.

38. Reid, *Hilbert*, 127.

39. Einstein et al., *Collected Papers*, 109.

40. Ibid., 540.

41. Ibid., 547.

42. Ibid., 568.

43. Ibid., 111.

44. Rowe, "Einstein Meets Hilbert."

45. Einstein et al., *Collected Papers*, 150.

46. Ibid., 154.

47. Phillips, "General Relativity."

48. Brandon Specktor, "Rare Einstein Manuscript Sells for Record-Smashing $13 Million at Auction," *Live Science*, July 28, 2022, www.livescience.com/albert-einstein-record-selling-manuscript-relativity.

49. Constance Reid, *Hilbert* (Berlin; Heidelberg: Springer Verlag, 1970), 142.

50. Renate Tobies, *Felix Klein: Visions for Mathematics, Applications, and Education*, trans. Valentine A. Pakis (Boston: Birkhäuser, 2021), 540.

51. Ibid., 541.

52. Ibid., 576.

53. Ibid., 540.

54. John D. Norton, "A Peek into Einstein's Zurich Notebook," 2012, https://sites.pitt.edu/~jdnorton/Goodies/Zurich_Notebook/.

55. Dennis Overbye, "A Century Ago, Einstein's Theory of Relativity Changed Everything," *New York Times*, November 24, 2015, www.nytimes.com/2015/11/24/science/a-century-ago-einsteins-theory-of-relativity-changed-everything.html.

56. Pais, *Subtle Is the Lord*, 167.

57. Amanda Gefter, "Newton's Apple: The Real Story," *New Scientist*, January 18, 2010, www.newscientist.com/article/2170052-newtons-apple-the-real-story/.

58. Einstein, "Maxwell's Influence."

59. Einstein et al., *Collected Papers*, 445.

60. Walter Isaacson, "How Einstein Reinvented Reality," *Scientific American*, September 2015.

61. Einstein et al., *Collected Papers*, 265.

62. Matthew Stanley, *Einstein's War: How Relativity Triumphed amid the Vicious Nationalism of World War I* (New York: Dutton, 2019).

63. Einstein et al., *Collected Papers*, 279.

64. Rowe, "Einstein Meets Hilbert," 408.

65. Pais, *Subtle Is the Lord*, 275.

66. Einstein et al., *Collected Papers*, 315.

67. Ibid., 320.

68. Ibid., 282.

69. Ibid., 292.

70. Albert Einstein, "What Is the Theory of Relativity?," *London Times*, November 28, 1919, http://germanhistorydocs.ghi-dc.org/pdf/eng/EDU_Einstein_ENGLISH.pdf.

71. Thomas Levenson, *The Hunt for Vulcan: . . . and How Albert Einstein Destroyed a Planet, Discovered Relativity, and Deciphered the Universe* (New York: Random House, 2015).

72. Michel Janssen and Jürgen Renn, "History: Einstein Was No Lone Genius," *Nature* 527 (November 16, 2015): 298–300, www.nature.com/news/history-einstein-was-no-lone-genius-1.18793.

73. Einstein et al., *Collected Papers*, 159.

74. Ibid., 160. Emphasis in original.

75. Ibid., 194.

76. Ibid., 300.

Chapter 3: The Theorem

1. Emmy Noether, "Invariante Variationsprobleme," *Nachrichten von Der Königlichen Gesellschaft Der Wissenschaften Zu Göttingen, Mathematisch-Physicalische Klasse*, 1918, 235–257.

2. Cornelius Lanczos, *The Variational Principles of Mechanics* (Toronto: University of Toronto Press, 1949), xii.

3. Frank Wilczek, *A Beautiful Question: Finding Nature's Deep Design* (New York: Penguin Press, 2015), 280.

4. Kate Zernike, *The Exceptions: Nancy Hopkins, MIT, and the Fight for Women in Science* (New York: Simon & Schuster, 2023), 204.

5. Danielle Sedbrook, "Must the Molecules of Life Always Be Left-Handed or Right-Handed?," *Smithsonian Magazine*, July 28, 2016, www.smithsonianmag.com/space/must-all-molecules-life-be-left-handed-or-right-handed-180959956/.

6. "What Is Antimatter?" *New Scientist*, n.d., www.newscientist.com/definition/antimatter/.

7. Cory Mitchell, "Redenomination: Examples of Currency Rebalancing," *Investopedia*, May 23, 2022, www.investopedia.com/terms/r/redenomination.asp.

8. Kevin Jensen, *The Physics of Shakespeare*, manuscript in preparation.

9. William Shakespeare, Sonnet 154, line 14, Open Source Shakespeare, George Mason University, www.opensourceshakespeare.org/views/sonnets/sonnet_view.php?Sonnet=154.

10. Nina Byers, "E. Noether's Discovery of the Deep Connection Between Symmetries and Conservation Laws," July 16, 1998, Physics Department, University of California Los Angeles, http://arxiv.org/pdf/physics/9807044.pdf.

11. Kat Jivkova, "Emmy Noether's Breakthrough: Mathematical Symmetries Are Equivalent to Physical Conservation Laws," *Retrospect Journal*, June 6, 2021, https://retrospectjournal.com/2021/06/06/emmy-noethers-breakthrough-mathematical-symmetries-are-equivalent-to-physical-conservation-laws/.

12. Yvette Kosmann-Schwarzbach, *The Noether Theorems: Invariance and Conservation Laws in the Twentieth Century* (New York: Springer, 2010), 121.

13. Ibid., 85.

14. Robert M. Wald, *General Relativity* (Chicago: University of Chicago Press, 1984), 62.

15. J. Adam, L. Adamczyk, J. R. Adams, et al. (STAR Collaboration), "Measurement of e⁺e⁻ Momentum and Angular Distributions from Linearly Polarized Photon Collisions," *Physical Review Letters* 127, no. 21698 (2021), doi:10.1103/PhysRevLett.127.052302.

16. Murat Gunaydin, Edward Witten, and Cihan Saclioglu, "Feza Gürsey's Biography," Feza Gürsey Center for Physics and Mathematics, Boğaziçi University, n.d., https://fezagursey.bogazici.edu.tr/en/feza-gurseys-biography.

17. Emmy Noether, *Gesammelte Abhandlungen—Collected Papers* (English and German edition), ed. Nathan Jacobson (Berlin; New York: Springer, 1983; reprint 2013).

18. James Clerk Maxwell, "A Dynamical Theory of the Electromagnetic Field," *Philosophical Transactions of the Royal Society of London* 155 (1865): 486, doi:10.1098/rstl.1865.0008.

19. LIGO (Laser Interferometer Gravitational-Wave Observatory), home page, https://ligo.caltech.edu.

20. Dennis Overbye, "Gravitational Waves Detected, Confirming Einstein's Theory," *New York Times*, 2016, www.nytimes.com/2016/02/12/science/ligo-gravitational-waves-black-holes-einstein.html.

21. David Darling, "Hulse-Taylor Pulsar (PSR 1913+16)," 2016, www.daviddarling.info/encyclopedia/H/HulseTaylor.html.

22. Paul Halpern, "How Richard Feynman Convinced the Naysayers 60 Years Ago That Gravitational Waves Are Real," *Forbes*, March 7, 2017, www.forbes.com/sites/startswithabang/2017/03/07/how-richard-feynman-convinced-the-naysayers-that-gravitational-waves-were-real-60-years-ago/.

23. Richard Feynman, "The Laws of Induction," chap. 17 in *The Feynman Lectures on Physics*, vol. 2, ed. Michael A. Gottlieb and Rudolf Pfeiffer (Pasadena: California Institute of Technology, 1964, 2006, 2013), www.feynmanlectures.caltech.edu/II_17.html.

24. Yvette Kosmann-Schwarzbach, "Emmy Noether's Wonderful Theorem," *Physics Today* 64, no. 9 (2011): 62, doi:10.1063/PT.3.1263.

25. Sean Carroll, "Energy Is Not Conserved," February 22, 2010, www.preposterousuniverse.com/blog/2010/02/22/energy-is-not-conserved; Michael Weiss and John Baez, "Is Energy Conserved in General Relativity?," *Physics and Relativity FAQ*, Department of Mathematics, University of California Riverside, 2017, http://math.ucr.edu/home/baez/physics/Relativity/GR/energy_gr.html.

26. Richard Morris, *The Edges of Science: Crossing the Boundary from Physics to Metaphysics* (New York: Touchstone, 1990), 69.

27. Wald, *General Relativity*, 84.

28. Morris, *Edges of Science*, 181.

29. Ryuzo Sato and Rama V. Ramachandran, *Symmetry and Economic Invariance* (New York: Springer, 2014).

30. Stuart Feldman, "A Conversation with Alan Kay," *ACM Queue* 2, no. 9 (2004), https://queue.acm.org/detail.cfm?id=1039523.

31. David E. Rowe, "'Jewish Mathematics' at Göttingen in the Era of Felix Klein," *Isis* 77 (1986): 422–449, https://faculty.math.illinois.edu/~reznick/davidrowe.pdf.

32. Albert Einstein et al., *The Collected Papers of Albert Einstein*, vol. 8, *The Berlin Years: Correspondence, 1914–1918* (English translation supplement), trans. Ann M. Hentschel (Princeton, NJ: Princeton University Press, 1998), 265, http://einstein papers.press.princeton.edu/vol8-trans/. Emphasis in original.

33. TehPhysicalist, "Feynman: 'Greek' Versus 'Babylonian' Mathematics," video, YouTube, May 17, 2012, www.youtube.com/watch?v=YaUlqXRPMmY.

34. David W. Lewis, "David Hilbert and the Theory of Algebraic Invariants," *Irish Mathematical Society Bulletin* 33 (1994): 42–54, www.maths.tcd.ie/pub/ims/bull33/bull33_42-54.pdf.

35. Wilczek, *Beautiful Question*, 294.

36. Einstein et al., *Collected Papers*, 216.

37. Ibid., 568.

38. Ibid., 569.

39. Abraham Pais, *Subtle Is the Lord: The Science and the Life of Albert Einstein* (Oxford, UK: Oxford University Press, 1982), 205.

40. Sharon Bertsch McGrayne, *Nobel Prize Women in Science: Their Lives, Struggles, and Momentous Discoveries*, 2nd ed. (Washington, DC: Joseph Henry Press, 2001), 74.

41. Norbert Schappacher and Cordula Tollmien, "Emmy Noether, Hermann Weyl, and the Göttingen Academy: A Marginal Note," *Historia Mathematica* 43, no. 2 (2016): 194–197, doi:/10.1016/j.hm.2015.11.002.

Chapter 4: Emmy Noether Gets a Job

1. Emiliana P. Noether and Gottfried E. Noether, "Emmy Noether in Erlangen and Göttingen," in *Emmy Noether in Bryn Mawr*, ed. Bhama Srinivasan and Judith Sally (New York: Springer, 2011), 133–137.

2. David E. Rowe and Mechthild Koreuber, *Proving It Her Way* (Cham, Switzerland: Springer, 2020), 71.

3. Ibid., 75.

4. Sharon Bertsch McGrayne, *Nobel Prize Women in Science: Their Lives, Struggles, and Momentous Discoveries*, 2nd ed. (Washington, DC: Joseph Henry Press, 2001), 72.

5. Rowe and Koreuber, *Proving It Her Way*, 76.

6. Ibid., 82.

7. Emmy Noether, "Invariante Variationsprobleme," *Nachrichten von Der Königlichen Gesellschaft Der Wissenschaften Zu Göttingen, Mathematisch-Physicalische Klasse*, 1918, 235–257.

8. Ruth Edith Hagengruber, project director, "Edith Stein," History of Women Philosophers and Scientists, Paderborn University, Paderborn, Germany, 2023, https://historyofwomenphilosophers.org/projects/project/edith-stein/.

9. Renate Tobies, *Felix Klein: Visions for Mathematics, Applications, and Education*, trans. Valentine A. Pakis (Boston: Birkhäuser, 2021), 591.

10. Fred Jerome and Rodger Taylor, *Einstein on Race and Racism* (New Brunswick, NJ: Rutgers University Press, 2005).

11. Larry Riddle, "Grace Chisholm Young," Agnes Scott College, last updated January 15, 2022, www.agnesscott.edu/lriddle/women/young.htm.

12. C. Jones, *Femininity, Mathematics and Science, 1880–1914* (Palgrave Macmillan, 2009), 4. Note that some of the biographical notes about Emmy Noether in this source are inaccurate.

13. Ibid., 1.

14. Auguste Dick, *Emmy Noether: 1882–1935* (Boston: Birkhäuser, 1981), 50.

15. McGrayne, *Nobel Prize Women in Science*, 74.

16. Dick, *Emmy Noether: 1882–1935*, 50.

17. Ibid., 51.

18. Matthew Stanley, *Einstein's War: How Relativity Triumphed amid the Vicious Nationalism of World War I* (New York: Dutton, 2019), 181.

19. Dick, *Emmy Noether: 1882–1935*, 40.

20. David E. Rowe, *Emmy Noether: Mathematician Extraordinaire* (Cham, Switzerland: Springer, 2020), 229.

21. Rowe and Koreuber, *Proving It Her Way*, 90.

22. B. L. van der Waerden, "Der Multiplizitätsbegriff Der Algebraischen Geometrie," *Mathematische Annalen* 97 (1927): 756–774, https://zbmath.org/53.0103.02.

23. Elise Crull and Guido Bacciagaluppi, eds., *Grete Hermann: Between Physics and Philosophy* (Dordrecht, Netherlands: Springer, 2016), 4.

24. Ibid., 180.

25. Ibid., 7.

26. Lee Phillips, "A Brief History of Quantum Alternatives," *Ars Technica*, July 28, 2017, https://arstechnica.com/science/2017/07/a-brief-history-of-quantum -alternatives/.

27. William Poundstone, "John von Neumann: European Career, 1921–30," *Encyclopaedia Britannica*, 2023, www.britannica.com/biography/John-von-Neumann /Princeton-1930-42.

28. Kate Zernike, *The Exceptions: Nancy Hopkins, MIT, and the Fight for Women in Science* (New York: Simon & Schuster, 2023).

Chapter 5: The Purge

1. Hans-Joachim Dahms, "Appointment Politics and the Rise of Modern Theoretical Physics at Göttingen," in *Göttingen and the Development of the Natural Sciences*, ed. Nicolaas Rupke (Göttingen, Germany: Wallstein, 2002), 143–157.

2. Renate Tobies, *Felix Klein: Visions for Mathematics, Applications, and Education*, trans. Valentine A. Pakis (Boston: Birkhäuser, 2021), 160.

3. Ibid., 49.

4. History Working Group, Institute for Advanced Study, "Emmy Noether's Paradise," spring 2017, www.ias.edu/ideas/2017/emmy-noether's-paradise.

5. David E. Rowe, "'Jewish Mathematics' at Göttingen in the Era of Felix Klein," *Isis* 77 (1986): 422–449, https://faculty.math.illinois.edu/~reznick/davidrowe.pdf.

6. Leopold Infeld, *Quest: An Autobiography* (Providence, RI: American Mathematical Society, Chelsea Publishing, 2006).

7. Max Born, Hedwig Born, and Albert Einstein, *The Born–Einstein Letters*, trans. Irene Born (New York: Macmillan, 1971), 35.

8. Kanishka Singh, "Over 70% of US Jewish College Students Exposed to Antisemitism This School Year, Survey Finds," *Reuters*, 2023, www.reuters.com /world/us/over-70-us-jewish-college-students-exposed-antisemitism-this-school-year -survey-2023-11-29/.

9. David E. Rowe and Mechthild Koreuber, *Proving It Her Way* (Cham, Switzerland: Springer, 2020), 195.

10. Constance Reid, *Courant* (Göttingen, Germany; New York: Springer-Verlag, 1976; with new foreword, New York: Copernicus, 1996), 179.

11. Rowe, "'Jewish Mathematics' at Göttingen."

12. Helene van Rossum, "Being Jewish at Princeton: From F. Scott Fitzgerald's Days to the Center of Jewish Life," *The Reel Mudd*, Mudd Manuscript Library, Princeton University, April 29, 2011, https://blogs.princeton.edu/reelmudd/2011/04 /printecons-jewish-students-from-the-days-of-scott-fitzgerald-to-the-center-of -jewish-life/.

13. Constance Reid, *Hilbert* (Berlin; Heidelberg: Springer Verlag, 1970), 188.

14. Sanford L. Segal, *Mathematicians Under the Nazis* (Princeton, NJ: Princeton University Press, 2014), 372.

15. Michio Kaku, *Einstein's Cosmos: How Albert Einstein's Vision Transformed Our Understanding of Space and Time* (New York: W. W. Norton, 2004), 179.

16. Abraham Pais, *Subtle Is the Lord: The Science and the Life of Albert Einstein* (Oxford, UK: Oxford University Press, 1982), 185.

17. Kaku, *Einstein's Cosmos*, 178.

18. Born, Born, and Einstein, *Born–Einstein Letters*, v.

19. Auguste Dick, *Emmy Noether: 1882–1935* (Boston: Birkhäuser, 1981), 167.

20. Segal, *Mathematicians Under the Nazis*, 131.

21. Ibid.

22. Ibid., 143.

23. Dick, *Emmy Noether: 1882–1935*, 132.

24. Segal, *Mathematicians Under the Nazis*, 60.

25. Bernhard Rust, quoted in Maxine Block and E. Mary Trow, eds., *Current Biography, 1942* (New York: The H. W. Wilson Company, 1942), 727.

Chapter 6: Emmy Noether in America

1. Auguste Dick, *Emmy Noether: 1882–1935* (Boston: Birkhäuser, 1981), 76.

2. Isabella Löhr, "Emergency Committee in Aid of Displaced Foreign Scholars," *Transatlantic Perspectives*, German Historical Institute, updated July 29, 2018, www.transatlanticperspectives.org/entries/emergency-committee-in-aid-of-displaced-foreign-scholars/.

3. Sharon Bertsch McGrayne, *Nobel Prize Women in Science: Their Lives, Struggles, and Momentous Discoveries*, 2nd ed. (Washington, DC: Joseph Henry Press, 2001), 84.

4. Karen Hunger Parshall, "Training Women in Mathematical Research: The First Fifty Years of Bryn Mawr College (1885–1935)," *Mathematical Intelligencer* 37 (May 20, 2015): 71–83, doi:10.1007/s00283-015-9540-2.

5. McGrayne, *Nobel Prize Women in Science*, 87.

6. Reinhard Siegmund-Schultze, *Mathematicians Fleeing from Nazi Germany: Individual Fates and Global Impact* (Princeton, NJ: Princeton University Press, 2009), 214.

7. Ibid., 250.

8. Qinna Shen, "A Refugee Scholar from Nazi Germany: Emmy Noether and Bryn Mawr College," *Mathematical Intelligencer* 41, no. 3 (2019): 1–14, https://repository.brynmawr.edu/cgi/viewcontent.cgi?article=1019&context=german_pubs.

9. Dick, *Emmy Noether: 1882–1935*, 76.

10. "Emmy Noether at Bryn Mawr College," ArcGIS StoryMaps, n.d., https://storymaps.arcgis.com/stories/de8b145587294523bfd8c40df9b6d446.

11. David E. Rowe and Mechthild Koreuber, *Proving It Her Way* (Cham, Switzerland: Springer, 2020), 176.

12. Bryn Mawr College, "History and Legacies Overview," 2023, www.brynmawr.edu/about-college/history.

13. Parshall, "Training Women in Mathematical Research."

14. Ibid.

15. "Special Collections, Bryn Mawr Library" (Bryn Mawr, PA: Bryn Mawr College; Special Collections, n.d.).

16. Parshall, "Training Women in Mathematical Research."

17. Ibid.

18. David E. Rowe, "'Jewish Mathematics' at Göttingen in the Era of Felix Klein," *Isis* 77 (1986): 422–449, https://faculty.math.illinois.edu/~reznick/davidrowe.pdf.

19. Rowe and Koreuber, *Proving It Her Way*, 174.

20. "2012 Award for an Exemplary Program or Achievement in a Mathematics Department," *Notices of the AMS*, May 2012, www.ams.org/notices/201205/rtx120500667p.pdf.

21. Bhama Srinivasan and Judith Sally, eds., *Emmy Noether in Bryn Mawr* (New York: Springer, 2011), 144.

22. Shen, "Refugee Scholar from Nazi Germany."

23. David E. Rowe, *Emmy Noether: Mathematician Extraordinaire* (Cham, Switzerland: Springer, 2020), 221.

24. Grace S. Quinn, Ruth S. McKee, Marguerite Lehr, and Olga Taussky, "Emmy Noether in Bryn Mawr," in *Emmy Noether in Bryn Mawr*, ed. Bhama Srinivasan and Judith Sally (New York: Springer, 2011), 139–146.

25. Leopold Infeld, *Quest: An Autobiography* (Providence, RI: American Mathematical Society, Chelsea Publishing, 2006).

26. Shen, "Refugee Scholar from Nazi Germany."

27. Rowe and Koreuber, *Proving It Her Way*, 192.

28. Quinn et al., "Emmy Noether in Bryn Mawr."

29. Shen, "Refugee Scholar from Nazi Germany."

30. Siegmund-Schultze, *Mathematicians Fleeing from Nazi Germany*, 244.

31. Shen, "Refugee Scholar from Nazi Germany."

32. Srinivasan and Sally, *Emmy Noether in Bryn Mawr*, 140.

33. *Bryn Mawr Alumnae Bulletin*, February 1935, Special Collections, Bryn Mawr Library, Bryn Mawr College, PA.

34. Special Collections, Bryn Mawr Library, Bryn Mawr College, PA. Hereafter cited as Special Collections, Bryn Mawr Library.

35. Wolfgang Saxon, "Dr. Elizabeth Monroe Boggs, 82, Founder of Group for Retarded," *New York Times*, January 30, 1996, www.nytimes.com/1996/01/30/nyregion /dr-elizabeth-monroe-boggs-82-founder-of-group-for-retarded.html.

36. B. L. van der Waerden, "The School of Hilbert and Emmy Noether," *Bulletin of the London Mathematics Society* 15, no. 1 (1983): 1–7, doi:10.1112/blms/15.1.1.

37. Saunders Mac Lane, "Van der Waerden's Modern Algebra," *Notices of the AMS* 44, no. 3 (1997): 321–322, www.ams.org/notices/199703/maclane.pdf.

38. Dick, *Emmy Noether: 1882–1935*, 165.

39. Ibid., 131.

40. Van der Waerden, "School of Hilbert and Emmy Noether."

41. Rowe, *Emmy Noether: Mathematician Extraordinaire*, 241.

42. Srinivasan and Sally, *Emmy Noether in Bryn Mawr*, 141.

43. Helene van Rossum, "Being Jewish at Princeton: From F. Scott Fitzgerald's Days to the Center of Jewish Life," *The Reel Mudd*, Mudd Manuscript Library, Princeton University, April 29, 2011, https://blogs.princeton.edu/reelmudd/2011/04 /printecons-jewish-students-from-the-days-of-scott-fitzgerald-to-the-center-of -jewish-life/.

44. Ron Unz, "Statistics Indicate an Ivy League Asian Quota," *New York Times*, updated December 3, 2013, www.nytimes.com/roomfordebate/2012/12/19/fears -of-an-asian-quota-in-the-ivy-league/statistics-indicate-an-ivy-league-asian-quota.

45. George Dyson, "Helen Dukas: Einstein's Compass," in *My Einstein Essays by Twenty-Four of the World's Leading Thinkers on the Man, His Work, and His Legacy*, ed. John Brockman (New York: Pantheon, 2006).

46. Constance Reid, *Courant* (Göttingen, Germany; New York: Springer-Verlag, 1976; with new foreword, New York: Copernicus, 1996), 206. The quoted remark is by Charles DePrima, who received his PhD from New York University in 1943 under Courant and taught at Caltech for forty years.

47. Richard Feynman, *Surely You're Joking, Mr. Feynman! (Adventures of a Curious Character)* (New York: W. W. Norton, 1997).

48. Dick, *Emmy Noether: 1882–1935*, 82.

49. Rowe and Koreuber, *Proving It Her Way*, 7.

50. Ibid., 25.

51. Srinivasan and Sally, *Emmy Noether in Bryn Mawr*, 139.

52. Rowe and Koreuber, *Proving It Her Way*, 140.

53. Ibid., 245.

54. Ibid.

55. Rowe, *Emmy Noether: Mathematician Extraordinaire*, 245.

56. Srinivasan and Sally, *Emmy Noether in Bryn Mawr*, 141.

57. Ibid., 143.

58. Ibid., 140 and 139.

59. Ibid., 142.

60. Ibid.

61. Rowe and Koreuber, *Proving It Her Way*, 189.

62. Ibid., 193.

63. Rowe, *Emmy Noether: Mathematician Extraordinaire*, 223.

64. Colin McLarty, "Poor Taste as a Bright Character Trait: Emmy Noether and the Independent Social Democratic Party," *Science in Context* 18, no. 3 (2005): 429–450, doi:10.1017/S0269889705000608.

65. Rowe and Koreuber, *Proving It Her Way*, 109.

66. McGrayne, *Nobel Prize Women in Science*, 88.

67. Special Collections, Bryn Mawr Library.

68. Quinn et al., "Emmy Noether in Bryn Mawr."

69. Rowe, *Emmy Noether: Mathematician Extraordinaire*, 267.

70. Srinivasan and Sally, *Emmy Noether in Bryn Mawr*, 143.

71. Ibid., 144.

72. Leon M. Lederman and Christopher T. Hill, *Symmetry and the Beautiful Universe* (Amherst, NY: Prometheus Books, 2004), 292.

73. Dick, *Emmy Noether: 1882–1935*.

74. Ibid., 153.

75. Albert Einstein, "The Late Emmy Noether," letter to the editor, *New York Times*, May 4, 1935, https://mathwomen.agnesscott.org/women/EinsteinNYTLetter.pdf.

76. Matthew Stanley, *Einstein's War: How Relativity Triumphed amid the Vicious Nationalism of World War I* (New York: Dutton, 2019), 158.

77. Rowe, "'Jewish Mathematics' at Göttingen in the Era of Felix Klein."

78. Einstein, "The Late Emmy Noether."

79. "Florence R. Sabin," *Profiles in Science*, National Library of Medicine, National Institutes of Health, n.d., https://profiles.nlm.nih.gov/spotlight/rr/feature/biographical-overview.

80. Special Collections, Bryn Mawr Library.

81. Ibid.

82. Burton Alan Boxerman, "Lucius Nathan Littauer," *American Jewish Historical Quarterly* 66, no. 4 (1977): 498–512.

83. "Six Months for Ex-Cong. Littauer," *Day* (London), February 4, 1914, 1, https://news.google.com/newspapers?nid=1915&dat=19140204&id=gPsgAAAAI BAJ&sjid=d3UFAAAAIBAJ&pg=1832,3535523.

84. "Littauer Judaic Book Fund," Bryn Mawr College, "Endowed Library Funds K–L," n.d., www.brynmawr.edu/inside/offices-services/library-information -technology-services/libraries-collections/endowed-library-funds/endowed-library -funds-k-l.

85. Special Collections, Bryn Mawr Library.

86. Ibid.

87. "Skinny Dipping in the Cloisters," *Beyond Bryn Mawr*, March 8, 2011, https:// beyondbrynmawr.wordpress.com/2011/03/08/skinny-dipping-in-the-cloisters/.

Chapter 7: Reawakening

1. Gerald Gabrielse, "The Standard Model's Greatest Triumph," *Physics Today* 66, no. 12 (2013): 64, doi:10.1063/PT.3.2223.

2. Saïda Guellati-Khelifa, "Searching for New Physics with the Electron's Magnetic Moment," *Physics* 16, no. 22 (2023), doi:10.1103/Physics.16.22.

3. Pierre Touboul, Gilles Métris, Manuel Rodrigues, et al., "MICROSCOPE Mission: Final Results of the Test of the Equivalence Principle," *Physical Review Letters* 129, no. 12 (September 2022), doi:10.1103/PhysRevLett.129.121102.

4. Albert Einstein et al., *The Collected Papers of Albert Einstein*, vol. 8, *The Berlin Years: Correspondence, 1914–1918* (English translation supplement), trans. Ann M. Hentschel (Princeton, NJ: Princeton University Press, 1998), 568, http://einstein papers.press.princeton.edu/vol8-trans/.

5. Chris Quigg, "Electroweak Symmetry Breaking in Historical Perspective," *Annual Review of Nuclear and Particle Science* 65 (2015): 25–42, doi:10.1146/annurev -nucl-102313-025537.

6. Hermann Weyl, *Space-Time-Matter*, trans. Henry L. Brose (London: Methuen, 1922).

7. Emmy Noether, "Invariante Variationsprobleme," *Nachrichten von Der Königlichen Gesellschaft Der Wissenschaften Zu Göttingen, Mathematisch-Physicalische Klasse*, 1918, 235–257.

8. Uta C. Merzbach, "Emmy Noether: Historical Contexts," in *Emmy Noether in Bryn Mawr*, ed. Bhama Srinivasan and Judith Sally (New York: Springer, 2011), 161–171.

9. J. Iliopoulos and J. R. Smith, "Plenary Report on Progress in Gauge Theories," in *High Energy Physics. Proceedings, 17th International Conference, ICHEP* (London, July 1, 1974), 89–116.

10. Yvette Kosmann-Schwarzbach, *The Noether Theorems: Invariance and Conservation Laws in the Twentieth Century* (New York: Springer, 2010), 64.

11. Nina Byers, "E. Noether's Discovery of the Deep Connection Between Symmetries and Conservation Laws" (presented at Symposium on the Heritage of Emmy Noether, Bar-Ilan University, Israel, December 2–4, 1996), http://arxiv.org /pdf/physics/9807044.pdf.

12. Peter Woit, *Not Even Wrong: The Failure of String Theory and the Search for Unity in Physical Law* (New York: Basic Books, 2006).

13. Sheldon Glashow, "Partial-Symmetries of Weak Interactions," *Nuclear Physics* 22, no. 4 (1961): 579–588, doi:10.1016/0029-5582(61)90469-2.

14. Sheldon Lee Glashow, "The Yang–Mills Model," *Inference* 5, no. 2 (2020), doi:10.37282/991819.20.1.

15. Noether, "Invariante Variationsprobleme."

16. Clark H. Kimberling, "Emmy Noether," *American Mathematical Monthly* 79, no. 2 (1972): 136–149, doi:10.1080/00029890.1972.11993006.

17. Renate Tobies, *Felix Klein: Visions for Mathematics, Applications, and Education*, trans. Valentine A. Pakis (Boston: Birkhäuser, 2021), 541.

18. Peder Olesen Larsen and Markus von Ins, "The Rate of Growth in Scientific Publication and the Decline in Coverage Provided by Science Citation Index," *Scientometrics* 83, no. 3 (2010): 575–603, doi:10.1007/s11192-010-0202-z.

19. R. M. F. Houtappel, H. Van Dam, and E. P. Wigner, "The Conceptual Basis and Use of the Geometric Invariance Principles," *Reviews of Modern Physics* 37 (1965): 595.

20. Eugene P. Wigner, "Events, Laws of Nature, and Invariance Principles," Nobel Lecture, December 12, 1963, Nobel Prize website, www.nobelprize.org/uploads/2018/06/wigner-lecture.pdf.

21. "The Nobel Prize in Physics 1963," Nobel Prize website, 1963, www.nobelprize.org/prizes/physics/1963/summary/.

22. Eugene P. Wigner, "The Unreasonable Effectiveness of Mathematics in the Natural Sciences," *Communications on Pure and Applied Mathematics* 13 (1960): 1–14.

23. Morris Kline, *Mathematics and the Physical World* (New York: Thomas Y. Crowell, 1959), 464.

24. Bryce S. DeWitt, *Dynamical Theory of Groups and Fields* (New York: Gordon and Breach, 1965), 585–820.

25. Richard Courant and David Hilbert, *Methods of Mathematical Physics* (New York: Interscience Publishers, 1937).

26. Constance Reid, *Courant* (Göttingen, Germany; New York: Springer-Verlag, 1976; with new foreword, New York: Copernicus, 1996), 92.

27. Ibid., 164.

28. Ibid., 170.

29. E. L. Hill, "Hamilton's Principle and the Conservation Theorems of Mathematical Physics," *Reviews of Modern Physics* 23, no. 3 (1951): 253, doi:10.1103/RevModPhys.23.253.

30. Kosmann-Schwarzbach, *Noether Theorems*, 102.

31. Yvette Kosmann-Schwarzbach, "Emmy Noether's Wonderful Theorem," *Physics Today* 64, no. 9 (2011): 62, doi:10.1063/PT.3.1263.

32. Lee Phillips, "General Relativity: 100 Years of the Most Beautiful Theory Ever Created," *Ars Technica*, 2015, http://arstechnica.com/science/2015/12/general-relativity-100-years-of-the-most-beautiful-theory-ever-created/.

33. Jakob Schwichtenberg, *Physics from Symmetry* (Cham, Switzerland: Springer, 2018), 101.

34. Werner Heisenberg, *Physics and Beyond* (New York: Harper and Row, 1971), 133.

35. John F. Donoghue, Eugene Golowich, and Barry R. Holstein, *Dynamics of the Standard Model* (Cambridge, UK; New York: Cambridge University Press, 2023), 23.

36. Matthew Stanley, *Einstein's War: How Relativity Triumphed amid the Vicious Nationalism of World War I* (New York: Dutton, 2019), 255.

37. Ibid., 325.

Chapter 8: Legacy

1. Emiliana P. Noether and Gottfried E. Noether, "Emmy Noether in Erlangen and Göttingen," in *Emmy Noether in Bryn Mawr*, ed. Bhama Srinivasan and Judith Sally (New York: Springer, 2011), 134.

2. Ibid.

3. Yvette Kosmann-Schwarzbach, *The Noether Theorems: Invariance and Conservation Laws in the Twentieth Century* (New York: Springer, 2010).

4. Hilary Lamb, "Interview with Ian Stewart," *Times Higher Education*, January 26, 2017, www.timeshighereducation.com/people/interview-ian-stewart -university-of-warwick. Emphasis mine.

5. Israel Kleiner, *A History of Abstract Algebra* (Boston: Birkhäuser, 2007), 99 and 157.

6. Sharon Bertsch McGrayne, *Nobel Prize Women in Science: Their Lives, Struggles, and Momentous Discoveries*, 2nd ed. (Washington, DC: Joseph Henry Press, 2001), 73.

7. Uta C. Merzbach, "Emmy Noether: Historical Contexts," in *Emmy Noether in Bryn Mawr*, ed. Bhama Srinivasan and Judith Sally (New York: Springer, 2011), 161–171.

8. Peter Roquette, "Emmy Noether and Hermann Weyl" (presentation, Hermann Weyl Conference, Bielefeld, Germany, September 10, 2006), www.mathi.uni -heidelberg.de/~roquette/weyl+noether.pdf.

9. Merzbach, "Emmy Noether."

10. Auguste Dick, *Emmy Noether: 1882–1935* (Boston: Birkhäuser, 1981), 167.

11. Merzbach, "Emmy Noether."

12. McGrayne, *Nobel Prize Women in Science*, 76.

13. Kosmann-Schwarzbach, *Noether Theorems*.

14. Ruth Gregory, "Who Was Emmy Noether?" video, *Physics World*, October 5, 2015, http://physicsworld.com/cws/article/multimedia/2015/oct/05/who-was-emmy -noether.

15. Sean Carroll, "The Nobel Prize Is Really Annoying," October 7, 2013, www .preposterousuniverse.com/blog/2013/10/07/the-nobel-prize-is-really-annoying/.

16. Brian Koberlein, "Symmetry," March 22, 2014, https://briankoberlein.com /blog/symmetry/.

17. Brian Koberlein, "The Power of Balance," October 28, 2016, https://archive .briankoberlein.com/2016/10/28/the-power-of-balance/index.html.

18. Leon M. Lederman and Christopher T. Hill, *Symmetry and the Beautiful Universe* (Amherst, NY: Prometheus Books, 2004), 21. Emphasis added.

19. Sophie Hermann and Matthias Schmidt, "Variance of Fluctuations from Noether Invariance," *Communications Physics* 5 (2022): 276, doi:10.1038/s42005-022 -01046-3.

20. Sabine Hossenfelder, "We Don't Know How the Universe Began, and We Will Never Know," video (with transcript), *BackReAction*, August 27, 2022, http:// backreaction.blogspot.com/2022/08/we-dont-know-how-universe-began-and -we.html.

21. Salvatore Capozziello and Antonio De Felice, "F(R) Cosmology from Noether's Symmetry," *Journal of Cosmology and Astroparticle Physics* 2008, no. 8 (2008): 14, doi:10.1088/1475-7516/2008/08/016.

22. Sabine Hossenfelder, "What's Going Wrong in Particle Physics?," video, *BackReAction*, February 11, 2023, http://backreaction.blogspot.com/2023/02 /whats-going-wrong-in-particle-physics.html.

23. Katie Mack, "The Inventor of Abstract Algebra," *Cosmos*, October 19, 2015, https://cosmosmagazine.com/science/physics/the-woman-who-invented-abstract -algebra/.

24. Erik D. Fagerholm, W. M. C. Foulkes, Yasir Gallero-Salas, Fritjof Helmchen, Karl J. Friston, Rosalyn J. Moran, and Robert Leech, "Conservation Laws by Virtue of Scale Symmetries in Neural Systems," *PLOS Computational Biology*, May 4, 2020, doi:10.1371/journal.pcbi.1007865.

25. Vishalxviii, "Introduction to Quantum Computing," *GeeksforGeeks*, updated May 9, 2023, www.geeksforgeeks.org/introduction-quantum-computing/.

26. Evan M. Fortunato, "Controlling Open Quantum Systems" (PhD thesis, Massachusetts Institute of Technology, 2002), 56.

27. Ryuzo Sato and Rama V. Ramachandran, *Symmetry and Economic Invariance*, 2nd ed. (Tokyo: Springer, 2014).

28. R. Sato, T. Nôno, and F. Mimura, "Hidden Symmetries: Lie Groups and Economic Conservation Laws," in *Operations Research and Economic Theory*, ed. H. Hauptmann, W. Krelle, and K. C. Mosler (Berlin: Springer, 1984).

29. Bey-Ling Sha, "Noether's Theorem: The Science of Symmetry and the Law of Conservation," *Journal of Public Relations Research* 16, no. 4 (2004): 391–416, www .tandfonline.com/doi/abs/10.1207/s1532754xjprr1604_4.

30. Nicolaas Rupke, ed., *Göttingen and the Development of the Natural Sciences* (Göttingen, Germany: Wallstein, 2002).

31. Hans-Joachim Dahms, "Appointment Politics and the Rise of Modern Theoretical Physics at Göttingen," in *Göttingen and the Development of the Natural Sciences*, ed. Nicolaas Rupke (Göttingen, Germany: Wallstein, 2002), 143–157.

32. Laurence Young, *Mathematicians and Their Times: History of Mathematics and Mathematics of History* (Amsterdam; New York: North-Holland, 2012), 2.

Appendix

1. Ingrid Daubechies and Shannon Hughes, "Konigsberg Bridge Problem," in Math Alive: Graph Theory, Princeton University, last updated August 2008, http://web.math.princeton.edu/math_alive/5/Lab1/Konigsberg.html.

2. Constance Reid, *Hilbert* (Berlin; Heidelberg: Springer Verlag, 1970), 11.

3. Paul Lockhart, *A Mathematician's Lament* (New York: Bellevue Literary Press, 2009).

4. Reid, *Hilbert*, 15.

5. Ibid., 29.

6. Ibid., 34.

7. Ibid., 33.

8. Chris K. Caldwell, "Euclid's Proof of the Infinitude of Primes (c. 300 BC)," PrimePages, 2021, https://primes.utm.edu/notes/proofs/infinite/euclids.html.

9. "Anselm: Ontological Argument for God's Existence," *Internet Encyclopedia of Philosophy*, www.iep.utm.edu/ont-arg/.

10. Reid, *Hilbert*, 33.

11. Burton Dreben and Akihiro Kanamori, "Hilbert and Set Theory," *Synthese* 110 (January 1997): 125, www.researchgate.net/profile/Akihiro_Kanamori2 /publication/225246572_HILBERT_AND_SET_THEORY/links/540705530cf2c -48563b28a7d.pdf.

12. Ibid.

13. Auguste Dick, *Emmy Noether: 1882–1935* (Boston: Birkhäuser, 1981), 17.

14. Uta C. Merzbach, "Emmy Noether: Historical Contexts," in *Emmy Noether in Bryn Mawr*, ed. Bhama Srinivasan and Judith Sally (New York: Springer, 2011), 161–171.

15. Hubert Goenner, "On the Shoulders of Giants: A Brief History of Physics in Göttingen," Institute of Theoretical Physics, Göttingen University, 2009, www .uni-goettingen.de/en/266241.html.

16. David Hilbert, *The Foundations of Geometry*, trans. E. J. Townsend (LaSalle, IL: Open Court, 1902), 2.

17. David Joyce, ed., *Euclid's Elements*, Book 1, Department of Mathematics and Computer Science, Clark University, 1996, http://aleph0.clarku.edu/~djoyce/elements /bookI/bookI.html.

18. Hilbert, *Foundations of Geometry*, 2.

19. Ibid., 8.

20. J. J. O'Connor and E. F. Robertson, "John Playfair (1748–1819)," MacTutor, School of Mathematics and Statistics, University of St Andrews, Scotland, updated August 1999, https://mathshistory.st-andrews.ac.uk/Biographies/Playfair/.

21. Euclid, *Euclid's Elements of Geometry*, trans. Richard Fitzpatrick (Richard Fitzpatrick, 2007), 7, www.cs.umb.edu/~eb/370/euclid/EuclidBook1.pdf.

22. J. J. O'Connor and E. F. Robertson, "Marcel Grossmann (1878–1936)," Mac-Tutor, School of Mathematics and Statistics, University of St Andrews, Scotland, updated May 2010, https://mathshistory.st-andrews.ac.uk/Biographies/Grossmann/.

23. John Gribbin, *Einstein's Masterwork: 1915 and the General Theory of Relativity* (New York: Pegasus Books, 2017), 141.

24. Ibid., 142.

25. Ransom Stephens, "Emmy Noether and the Fabric of Reality," video, YouTube, June 16, 2010, www.youtube.com/watch?v=1_MpQG2xXVo&feature=youtu.be&a.

26. Morris Kline, *Mathematics and the Physical World* (New York: Thomas Y. Crowell, 1959), 443.

27. Morris Kline, *Mathematics in Western Culture* (New York: Oxford University Press, 1953), 425.

28. Richard Feynman, *The Character of Physical Law (Messenger Lectures, 1964)* (Cambridge, MA: MIT Press, 2001), 103.

29. Manfred R. Schroeder, "Number Theory and the Real World," *Mathematical Intelligencer* 7 (1985).

30. Cornelius Lanczos, *The Variational Principles of Mechanics* (Toronto: University of Toronto Press, 1949), 385.

31. Thomas Levenson, *The Hunt for Vulcan: . . . and How Albert Einstein Destroyed a Planet, Discovered Relativity, and Deciphered the Universe* (New York: Random House, 2015).

32. John Baez, "Mysteries of the Gravitational 2-Body Problem," Department of Mathematics, University of California Riverside, January 18, 2022, https://math.ucr.edu/home/baez/gravitational.html.

33. Herbert Goldstein, "More on the Prehistory of the Laplace or Runge–Lenz Vector," *American Journal of Physics* 44, no. 11 (1976): 1123, doi:10.1119/1.10202.

34. Dick, *Emmy Noether: 1882–1935*, 145.

35. Alberto Maria Bersani and Paolo Caressa, "Lagrangian Descriptions of Dissipative Systems: A Review," *Mathematics and Mechanics of Solids* 26, no. 6 (2021): 785–803, doi:10.1177/1081286520971834.

36. Association for Women in Mathematics, "Karen E. Smith Named 2016 Noether Lecturer," press release, August 3, 2015, http://pages.uoregon.edu/vitulli/PR_SmithNoether2016.pdf.

37. International Mathematical Union, "ICM Emmy Noether Lecture," n.d., www.mathunion.org/imu-awards/icm-emmy-noether-lecture.

38. Anthony Bonato, "Announcing the Emmy Noether Scholarship," *The Intrepid Mathematician*, April 12, 2017, https://anthonybonato.com/2017/04/12/announcing-the-emmy-noether-scholarship/.

39. European Physical Society, "EPS Emmy Noether Distinction for Women in Physics," 2022, www.eps.org/page/distinction_prize_en.

40. Institute of International Education Scholar Rescue Fund, "IIE-SRF Establishes the Emmy Noether Chair to Support Outstanding Women Scholars Facing Threats to Their Lives and Careers," n.d., www.scholarrescuefund.org

/news_and_events/iie-srf-establishes-the-emmy-noether-chair-to-support-outstanding
-women-scholars-facing-threats-to-their-lives-and-careers/.

41. Perimeter Institute, "Emmy Noether Initiatives," 2020, https://perimeter
institute.ca/emmy-noether-initiatives.

42. Brookhaven National Laboratory, Computational Science Initiative, "Noether Fellowship Eligibility," n.d., www.bnl.gov/compsci/noether/application.php.

43. Steven G. Avery and Burkhard U. W. Schwab, "Noether's Second Theorem and Ward Identities for Gauge Symmetries," *Journal of High Energy Physics* 2 (2016): 1–31, doi:10.1007/JHEP02%282016%29031.

44. Steven Weinberg, "What Is Quantum Field Theory, and What Did We Think It Is?" (talk, Historical and Philosophical Reflections on the Foundations of Quantum Field Theory, Boston University, March 1996), 1997, https://arxiv.org/pdf/hep-th/9702027.pdf.

45. Thomas A. Garrity, *Electricity and Magnetism for Mathematicians: A Guided Path from Maxwell's Equations to Yang-Mills* (New York: Cambridge University Press, 2015), 85.

46. Ali Sanayei and Otto E. Rössler, *Chaotic Harmony: A Dialog About Physics, Complexity and Life* (New York: Springer, 2014).

47. Emiliana P. Noether and Gottfried E. Noether, "Emmy Noether in Erlangen and Göttingen," in *Emmy Noether in Bryn Mawr*, ed. Bhama Srinivasan and Judith Sally (New York: Springer, 2011), 133–137.

48. For the first quotation, see Stephens, "Emmy Noether and the Fabric of Reality." For the second, see Ransom Stephens, "The Unrecognized Genius of Emmy Noether," n.d., https://tgp.ransomstephens.com/emmy.htm.

49. Ransom Stephens, *The God Patent* (San Rafael, CA: Vox Novus, 2009).

50. Stephen Rennicks, "Zwischenwelt Telekinesis Video, Noether Symmetry, and More," *Drexciya Research Lab*, January 10, 2016, http://drexciyaresearchlab.blog spot.ie/2016/01/zwischenwelt-telekinesis-video-noether.html.

51. Eugenia Cheng, "Having Your Pi and Eating It: Talking Maths with Eugenia Chen," Gonville & Caius College, University of Cambridge, March 17, 2017, www.cai .cam.ac.uk/news/having-your-pi-and-eating-it-talking-maths-eugenia-cheng; Sarah Guminski, "Emmy Noether's Mathematics as Hotel Decor," *Scientific American*, May 23, 2017, https://blogs.scientificamerican.com/observations/emmy-noethers -mathematics-as-hotel-decor/.

52. Randall Munroe, "Marie Curie," comic panel, *XKCD* (web comic), n.d., https://xkcd.com/896/; Randall Munroe, "Advanced Techniques," comic panel, *XKCD* (web comic), n.d., https://xkcd.com/2595/. You must hover over the panels with your mouse to see the bonus title text.

53. Lee Phillips, "A Brief History of Quantum Alternatives," *Ars Technica*, July 28, 2017, https://arstechnica.com/science/2017/07/a-brief-history-of-quantum-alternatives/.

54. Chris Lipa, Cornell Math Explorers' Club, "Conway's Game of Life," Cornell Math Explorers' Club, Cornell University, n.d., https://pi.math.cornell.edu/~lipa /mec/lesson6.html.

55. Lee Phillips, *Practical Julia: A Hands-On Introduction for Scientific Minds* (San Francisco: No Starch Press, 2023). For the version in a single line of APL, see DyalogLtd, "Conway's Game of Life in APL," video, YouTube, January 26, 2009, www.youtube.com/watch?v=a9xAKttWgP4.

56. U. Frisch, B. Hasslacher, and Y. Pomeau, "Lattice Gas Automata for the Navier-Stokes Equations," *Physical Review Letters* 56, no. 14 (1986): 1505–1508, http://new.math.uiuc.edu/im2008/dakkak/papers/files/fhp1986.pdf.

57. Reza Olfati-Saber, "Nonlinear Control of Underactuated Mechanical Systems with Application to Robotics and Aerospace Vehicles" (PhD thesis, Massachusetts Institute of Technology, 2001).

58. Jozef Hanc, Slavomir Tuleja, and Martina Hancova, "Symmetries and Conservation Laws: Consequences of Noether's Theorem," *American Journal of Physics*, 72, no. 4 (2004): 428–435, www.eftaylor.com/pub/symmetry.html.

Index

abstract algebra, 285–288, 286 (fig.)
  Noether, E., and, 63, 131, 189,
    244–245, 285
  symmetry in, 285–286
accelerating frames, 43
acceleration
  of electrons, 291–292
  in $F = ma$, 70–71
  gravity and, 47
Alexandrov, Pavel, 170, 188,
    205–206, 245
Althoff, Friedrich, 15–16, 20, 38
American Mathematical Society, 179,
    180, 297
American Men of Science (McKeen), 179
Ampère, André-Marie, 89–90
angular momentum, 107, 126–127
  conservation of, 257

  of electron, 219
  gravity and, 289–290, 289 (fig.)
  in Noether's theorem, 289
antimatter twins, 102–103, 252
antiprotons, 103
antisemitism, 164–174, 182
  Einstein and, 168–169, 208
  at Göttingen University, 166
  after World War I, 165
Aristotle, 273
Arnold, V. I., 308
associative property, 287
axioms (postulates), 61–62, 63

the Beatles, 211
A Beautiful Question (Wilczek), 139
Bell, John Stuart, 162
Besso, Michele, 82, 84, 96

Big Bang, 250, 262
  preponderance of matter over
      antimatter and, 103
  expanding universe and, 129
black holes, 96, 262
Blumenthal, Otto, 26, 36
Boggs, Elizabeth Monroe, 185–187,
      198–199
Bohr, Harald, 38
Bohr, Niels, 1
Bonato, Anthony, 297
Boole, George, 267
Boolean algebra, 267
Born, Max, 16, 96
  Einstein and, 165, 170
  at Göttingen University, 163
  Hilbert, D., and, 56
  Minkowski and, 56
  quantum mechanics and, 284
  representation theory and, 295
Brauer, Richard, 178–179
Bryn Mawr
  Cloisters at, 212 (photo), 213,
      214–215
  Noether, E., at, 178–202, 285

calculus
  for curved space, 67–68
  for curved space-time, 71–72
  differential, 72, 281–283
  tensor, 71–73
  variational, 131, 228, 233
Cattell, James McKeen, 179
cellular automata, 305
centrifugal force, 281

Chaotic Harmony (Sanayei and
      Rössler), 298–299
Cheng, Eugenia, 303
Chicago, 303
chirality, 192
classical mechanics
  conservation laws and, 6–7
  energy conservation in, 127
  Noether's theorem and, 215, 223,
      224, 225, 229, 249
  space-time in, 226
  symmetry in, 117–118, 226, 229
Cloisters, at Bryn Mawr, 212 (photo),
      213, 214–215
conservation laws. *See also* energy
      conservation
  classical mechanics and, 6–7
  Einstein and, 142
  gambler's fallacy and, 106–107
  gravity and, 290
  hidden, 257–258
  Lagrangian and, 223
  in Noether's theorem, 100–109, 120,
      129–130, 232, 257, 290
  in standard model, 232
  symmetry and, 6–7, 100–109, 120, 121,
      127, 129–131, 250–251, 283–284
constraints, 307
continuity equation, 282–283
control theory, 307–308
converses and Noether's theorem,
      108–109, 246
Conway's Game of Life, 303–306, 304
      (fig.), 306 (fig.)
Copenhagen interpretation, 301

Copernicus, Nicolaus, 44

Corry, Leo, 56, 57

cosmos/cosmology. *See also* Big Bang; expanding universe

  energy in, 128–129

  general theory of relativity in, 133

  in Noether's theorem, 127, 129–130, 248

  uniformity of substance and, 140

Courant, Richard, 151, 163, 166, 175

  at Göttingen University, 228, 229

  Hilbert, D., and, 228

  Methods of Mathematical Physics by, 228

  Nazis and, 228

  Taussky and, 192

covariance, 76–77

  general, 76, 96

  general theory of relativity and, 90–91

  gravity and, 83

  Hilbert, D., and, 83

crystallography, 189

curved space, 65–68

curved space-time

  calculus for, 71–72

  in general theory of relativity, 70, 122

  gravity and, 83

dark energy, 248

definitions, 60–61, 63

DeWitt, Bryce, 227

differential calculus, 72, 281–283

differential equations, 70–71

for energy conservation, 113

  in Newton's laws of motion, 117

Dirac, Paul, 1, 252

Doppler effect, 128

Dynamical Theory of Groups and Fields (summer school course by DeWitt), 227

eclipse, 238

econometrics, 257–258

Eddington, Arthur, 236–239

eigenfunctions, 284

Einstein, Albert, 8

  antisemitism and, 168–169, 208

  arrogance of, 66

  Born and, 165, 170

  conservation law and, 142

  curved space and, 67–68

  energy conservation and, 115, 140–144, 151

  formality and, 64–65

  general theory of relativity of, 2, 29, 33, 47–48, 70, 80–81, 83–86, 89–92, 94, 95, 96, 97, 231

  at Göttingen University, 53–97, 279

  gravity and, 45, 46–47, 116, 129, 131–132, 239, 288–294, 289 (fig.)

  Grossmann and, 67–68, 70, 80, 278–279

  Hilbert, D., and, 69–70, 77–83, 86, 91, 116, 131–134, 141–142, 231, 262

  hometown of, 16

  at Institute for Advanced Study, 177, 190, 191, 194–195

  inventions and patents of, 94

Einstein, Albert *(Continued)*
    Klein and, 73, 74, 143
    Mercury and, 82, 95–96, 97, 207, 251, 288, 294
    Minkowski and, 32–34
    miracle year of, 27–28, 66, 67
    Nobel Prize of, 28, 237, 246
    Noether, E., and, 11, 66–67, 74–76, 81, 150–152, 206–208, 247
    Noether's theorem and, 140–144, 242
    as patent clerk, 28–29, 48
    philosophy and, 93
    Poincaré and, 86–87
    principle of equivalence of, 46–48
    quantum mechanics and, 161, 237
    as schoolteacher, 28–29
    special theory of relativity of, 28, 29, 30–33, 43, 69
    statistical mechanics and, 249
    tensor calculus and, 71–73
    theory of relativity of, 29–30
    to US, 4
    variational principle and, 92–93
    Weyl and, 91–92, 133, 220
Einstein summation convention, 142
electrodynamics, 88–90, 114
electromagnetism, 30–31, 126–127
    energy conservation and, 110
    energy in, 123
    gauge theories and, 221, 222
    general theory of relativity and, 133
    gravity and, 90, 231
    Hilbert, D., and, 231
    Maxwell on, 31, 88–90, 222, 282, 285
    quantum mechanics and, 285

electrons, 103, 121, 134
    acceleration of, 291–292
    angular momentum of, 219
    magnetic moment of, 219
    uniformity of substance and, 139
    wave function of, 292
elementary particles. *See also specific types*
    antimatter twins of, 102–103, 252
    general theory of relativity silent on, 217
    mass of, 217
    Noether's theorem and, 2
    in standard model, 236
    symmetry and, 226, 236
    Wigner and, 226
    Wilczek and, 100, 139
Elements (Euclid), 61–63, 134
$E = mc^2$, 28, 67, 106, 121, 237
Emergency Committee in Aid of Displaced Foreign Scholars, 175, 176–177, 182, 297
Emmy Noether Fellowship, 181
Emmy Noether Memorial Fund, 208–211, 209 (photo)
Emmy Noether Scholarship, 297
energy
    in cosmos, 128–129
    defined, 122–123
    in electromagnetism, 123
    in $E = mc^2$, 28, 67, 106, 121, 237
    in expanding universe, 128
    in general theory of relativity, 129, 155

of gravitational waves, 125
levels in hydrogen, 292–293
kinetic, 111–113
Lorentz transformation and, 89
Maxwell and, 90
in Noether's theorem, 109–131
potential, 112–113
time translation symmetry and, 122
energy conservation, 109–131
in classical mechanics, 127
Einstein and, 140–144, 151
general theory of relativity and, 2,
114–115, 118, 128
global, 110, 114
gravity and, 130–131
Hilbert, D., and, 130–131, 141
ideal systems in, 111
isolated systems in, 110–111
Klein and, 141
in Noether's theorem, 117–118, 122,
127–131, 140
Shakespeare and, 113
space-time and, 124
symmetry and, 117–118
time translation symmetry and, 109
energy-momentum four vector,
121–122
Entwurf theory, 80
EPS. *See* European Physical Society
Euclid, 60–63, 134
Hilbert, D., and, 277–278, 278 (fig.),
279
on lines, 280
on prime numbers, 273
Euler, Leonhard, 266

European Physical Society (EPS),
297
The Evolution of Physics (Einstein and
Infeld), 165
Ewald, Paul, 37
expanding universe
Einstein and, 92
quantum mechanics and, 127–129

Faraday, Michael, 89–90
Fermat's last theorem, 68–69
Feynman, Richard, 126, 134, 283–284
on Institute for Advanced Study,
190–191
Noether's theorem and, 223
"Fields and Systems of Rational
Functions" (Noether, E.), 147
field theory, 87–89. *See also* quantum
field theory
econometrics and, 258
gauge theories, 222
general covariance of, 96
Fischer, Ernst, 34–35
Flexner, Abraham, 177, 190, 191
Flexner, Bill, 186
Flexner, Simon, 209
F = ma, 70–71
formalist mathematics, 274
formality, 64–65, 134
Förster, Rudolph, 91
Fortunato, Evan, 256
The Foundations of Geometry (Hilbert),
18, 41, 59, 62
Franco-Prussian War, 16–17
Frankfurt University, 149–150

Freund, Peter, 41
functions, 72

Galilean relativity, 29–30, 31
Galileo, Tower of Pisa experiment of,
46
gambler's fallacy, 106–107
Garrity, Thomas, 298
gauge theories
    described by Noether, 226
    of electromagnetism, 142, 221
    standard model and, 222
Gauss, Carl Friedrich, 41, 70
Gell-Mann, Murray, 223
general covariance, 76, 96
general theory of relativity
    cosmos in, 128–129, 133
    covariance and, 90–91
    curved space-time in, 70, 122
    Eddington and, 237–239
    of Einstein, 2, 29, 33, 47–48, 70,
        80–81, 83–86, 89–92, 231
    electromagnetism and, 133
    elementary particles and, 217
    energy conservation and, 2, 114–115,
        118, 128
    energy in, 129, 155
    energy-momentum four vector and,
        121–122
    Entwurf theory and, 80
    as field theory, 87, 89
    gauge theories and, 222
    gravitational mass in, 46, 219–220,
        251
    gravitational waves and, 124–125

gravity and, 43–44, 48, 121–124, 217,
        232, 238–239, 250, 288–289
    Grossmann and, 80–81, 84, 231
    Hilbert, D., and, 54–55, 75–76, 81–84,
        92, 116, 131, 133–134, 141–142, 152
    inertial mass in, 46, 219–220, 251
    Klein and, 82–86, 92, 132
    limiting condition in, 123–124
    local energy conservation and, 115,
        283
    Noether, E., and, 54–55, 75–76,
        84–86, 116–117, 131–135, 231, 242
    Noether's theorem and, 100,
        140–144, 217–235, 242, 289
    Poincaré and, 87
    principle of equivalence and, 47–48
    priority dispute, 83
    reference frames in, 43
    standard model and, 232
    symmetry in, 108
geodesics, 280
German Mathematicians' Relief Fund,
    195–196
Glashow, Sheldon, 222–223
global energy conservation, 110, 114
Gödel, Kurt, 59, 60
God particle, 3
God, ontological argument for
        existence of, 274
Goeppert Mayer, Maria, 226
Goldstein, Herbert, 224
Gorbachev, Mikhail, 201
Gordan, Paul, 148, 164
    Hilbert, D., and, 243–244, 272, 274
    invariant theory and, 275

Noether, E., and, 26–27, 34–36, 243
  obituary for, 42
Göttingen University
  antisemitism at, 163–174
  Born at, 163
  Bryn Mawr and, 179–180
  Courant at, 228, 229
  Einstein at, 53–97, 279
  gravity at, 53–97
  Hilbert, D., at, 19–20, 22–25, 36–43,
    53–97, 145–163, 169, 228, 260
  Klein at, 12, 14–20, 145–163
  meritocracy at, 12, 39, 147, 153
  Minkowski at, 32
  Noether, E., at, 21–22, 25–26,
    42–43, 53–97, 145–174,
    203–204, 259–260
  Poincaré at, 87
  Riemann at, 279
  spirit of place, 12
  Taussky at, 192
  Weyl at, 23–24, 203–204
grand unified theory
  particle physics and, 250–251
  standard model and, 232
graph theory, 266
gravitational lensing, 262
gravitational mass, 46, 219–220, 251, 281
gravitational waves, 115, 124–126, 220,
    262
gravity, 2
  angular momentum and, 289–290,
    289 (fig.)
  conservation laws and, 290
  covariance and, 76–77, 83

curved space and, 65–68
  curved space-time and, 83
  Einstein and, 45, 46–47, 53–97, 116,
    129, 131–132, 239, 278–281,
    288–294, 289 (fig.)
  electromagnetism and, 90, 231
  energy conservation and, 114–115,
    130–131
  field equations for, 131
  formality and, 64–65
  general theory of relativity and,
    43–44, 48, 121–124, 217, 232,
    238–239, 250, 288–289
  high-energy physics and, 249
  Hilbert, D., and, 53–97, 114–115,
    131–132, 231
  kinetic energy and, 113
  Klein and, 54, 114
  mass and, 46–47
  Maxwell and, 90
  Mercury and, 123–124
  Newton and, 44–46, 48, 65, 88,
    251
  Noether, E., and, 53–97
  Noether's theorem and, 140, 230,
    233, 238–239, 250
  potential energy and, 112–113
  quantum, 249
  space-time and, 232, 280–281
  standard model and, 218, 220, 233
  string theory and, 222
  symmetry and, 233
  variational principle and, 92–93
great circle, 280
Greene, Brian, 3–4

Gregory, Ruth, 246

Gribbin, John, 33

grid artifacts, 303

grids, 302–303, 302 (fig.)

Grossmann, Marcel, 29

  curved space and, 67–68, 70

  Einstein and, 67–68, 70, 80, 278–279

  general theory of relativity and, 54, 80–81, 84, 231

  space-time and, 278–279

group theory, 130, 188–189, 287–288, 290, 292–293

Group Theory and Quantum Mechanics (Weyl), 205

Gürsey, Feza, 122

habilitation of Noether, E., 145–163

"Hamilton's Principle and the Conservation Theorems of Mathematical Physics" (Hill, E.), 229–230

"Hamilton's Principle and the General Theory of Relativity" (Einstein), 141–142

Hartmann, Johannes, 148

Hasse, Helmut, 170–171

Hayes, Helen, 211

Heisenberg, Werner, 1

  matrix mechanics, meeting with Hilbert about, 284–285

  Noether's theorem and, 221, 233–234, 235

  particle physics and, 120, 223

  on the primacy of symmetry, 234

  representation theory and, 295

Hepburn, Katharine, 213

Hermann, Grete, 159–162

Herodotus, 9

Hertz, Paul, 69

hidden conservation laws, 257–258

hidden symmetries in Noether's theorem, 235–236, 289, 291

hidden variables in quantum mechanics, 161–162

Higgs, Peter, 223

Higgs boson, 248

high-energy physics. *See* particle physics

Hilbert, David, 8, 9, 10

  abstraction and, 40–41

  antisemitism and, 167–168

  Born and, 56

  Bryn Mawr and, 179

  celebrity, 36

  conflicts with Göttingen faculty, 149

  Courant and, 228

  covariance and, 83

  Einstein and, 69–70, 77–83, 86, 91, 116, 131–134, 141–142, 231, 262

  electromagnetism and, 231

  energy conservation and, 109–111, 116, 130–131, 141

  Euclid and, 62–63, 277–278, 278 (fig.), 279

  formality and, 64–65

  general theory of relativity and, 54–55, 75–76, 81–84, 92, 116, 131, 133–134, 141–142, 152

Gordan and, 243–244, 272, 274

at Göttingen University, 19–20, 22–25, 36–43, 53–97, 145–163, 169, 228, 260

gravity and, 53–97, 114–115, 131–132, 231

hometown of, 16

invariant theory and, 267, 272–274

Klein and, 12, 17–20, 23, 37–38, 92, 132, 147

at Königsberg, 18–20, 23, 267, 272

local energy conservation and, 283

"Manifesto to the Civilized World" and, 40, 79

meritocracy and, 153

Methods of Mathematical Physics by, 228

Minkowski and, 18, 19, 24, 25, 38, 65, 272

Nazis and, 173–174

Noether, E., and, 11, 26, 35, 42, 58–59, 75–76, 84, 122, 132, 143, 147–148, 149, 157, 193, 194, 242, 243

Noether's theorem and, 231, 242

peculiarities of, 19–20, 23, 24, 36–37

physics and, 55–60

political tactics of, 79

problems of, 25, 41

quantum mechanics and, 284–285

representation theory and, 295

retirement of, 169

Stark and, 165

statistical mechanics and, 65

unified theory and, 133–134, 231

as universalist, 134–135

variational principle of, 92–93

Weyl and, 220, 274

Hilbert, Franz (son), 24

Hilbert, Käthe Jerosch (wife), 24–25

Hilbert's basis theorem, 272–274

Hilbert space, 55, 134

Hill, Christopher, 3, 39, 202, 248

Hill, E. L., 229–230

history, 9

History of Abstract Algebra (Kleiner), 244

Hitler, Adolf, 171, 172

Hopf, Heinz, 245

Hossenfelder, Sabine, 251

Hubble, Edwin, 129

Hulse, Russell, 124–125

The Hunt for Vulcan (Levenson), 95

Hurwitz, Adolf, 19

hydrogen, 292–293

An Ideal Husband (Wilde), 259

ideal systems in energy conservation, 111

Iliopoulos, John, 221

inertial mass, 46, 219–220, 251, 281

inertial reference frames, 43

Infeld, Leopold, 165

instantaneous action at a distance, 48, 251

Institute for Advanced Study

Einstein at, 177, 190, 191, 194–195

Feynman on, 190–191

Noether, E., at, 189–192, 196–197, 202

Weyl at, 171–172

Institute of International Education, 297

integral equations, 134

International Congress of Mathematicians, 245

International Mathematical Union, 297

intuitionist mathematics, 274

invariant theory, 26, 267–275, 268 (fig.), 269 (fig.)
  Hilbert, D., and, 267, 272–274
  Minkowski and, 272
  Noether, E., and, 26, 29, 116, 143, 275
  special theory of relativity and, 29, 276

"Invariant Variational Problems" (Noether, E.), 99–100

Isaacson, Walter, 91

isolated systems in energy conservation, 110–111

Jensen, J. Hans D., 226

Jones, Claire G., 154

Joyce, James, 41

Kahn, Margarete, 153–154

Kandinsky, Wassily, 41

Kant, Emmanuel, 18

Kay, Alan, 131

Kepler, Johannes, 44, 291, 293

Kimberling, Clark H., 224

kinetic energy, 111–113

Klein, Felix, 8
  Althoff and, 15–16, 38
  antisemitism and, 167
  bottle of, 14

Bryn Mawr and, 179
  Einstein and, 64, 73, 74, 143
  energy conservation and, 115, 141
  general theory of relativity and, 82–86, 92, 132
  at Göttingen University, 12, 14–20, 145–163
  gravity and, 54, 114
  Hilbert, D., and, 12, 17–20, 23, 37–38, 92, 132, 147
  "Manifesto to the Civilized World" and, 40
  Noether, E., and, 26, 84–86, 115, 122, 143, 147–148, 157, 224–225
  Noether's theorem and, 242
  Planck and, 225
  at University of Erlangen, 164
  Young, G., and, 151, 262

Kleiner, Israel, 244

Koberlein, Brian, 247–248

Königsberg
  bridge problem, 18, 266
  Hilbert, D., at, 18–19, 23, 267, 272

Kosmann-Schwarzbach, Yvette, 222, 230, 246

Kovalevskaya, Sofya Vasilyevna, 153

Lagrangian, 222–223, 296, 298

Lanczos, Cornelius, 99

Landau, Edmund, 148, 166, 196

Landau, Lev, 47

Laplace, Pierre-Simon, 290

lattice gases, 305

Laue, Max von, 37–38

law of excluded middle, 273

Lectures on Physics (Feyman), 126

Lederman, Leon, 3, 39, 202, 248

Lefschetz, Solomon, 176–177, 184, 205

Lehr, Marguerite, 199

Leibniz, Gottfried Wilhelm, 131

Lenard, Philipp, 165

Lenin, Vladimir, 19, 197

Lenz, Emil, 290

Levi-Civita, Tullio, 71–72, 279

Levenson, Thomas, 95

liar's paradox, 59

Lie, Sophus, 189

Lie group theory, 130, 189, 290, 292–293, 296

LIGO experiment, 124

limiting condition, of special relativity, 123–124

Lindemann, Ferdinand von, 272

lines

  Euclid's axiom about, 61

  Euclid's definition of, 60–61

  generalized to geodesics, 280

Littauer, Lucius Nathan, 209–211

Löbenstein, Klara, 153–154

local energy conservation, 110, 114, 115, 281–283, 282 (fig.)

Lorentz transformation, 65, 88–89

Mack, Katie, 253

magnetic moment, 219

"Manifesto to the Civilized World," 40, 79

mass

  conservation law for, 106

  of elementary particles, 217

in $E = mc^2$, 28, 67, 106, 121, 237

in $F = ma$, 70–71

  gravitational and inertial, 46, 219–220, 251, 281

  gravity and, 46–47

  kinetic energy and, 111

  local conservation of, 283

Mathematical Methods of Classical Mechanics (Arnold), 308

Mathematische Annalen, 272

matrix mechanics, 285

Maxwell, James Clerk, 30–31, 123

  on electromagnetism, 31, 88–90, 222, 282, 285

  energy conservation and, 110

McGrayne, Sharon, 41, 244, 246

McKee, Ruth S., 182–183, 202

Mein Kampf (Hitler), 172

Mercury

  Einstein and, 82, 95–96, 97, 207, 251, 288, 294

  gravity and, 123–124

  perihelion, precession of, 44–45, 95–96, 207, 251, 288, 293–294

meritocracy

  at Göttingen University, 12, 39, 147, 153

  Hilbert, D., and, 153

Merzbach, Uta C., 245

Methods of Mathematical Physics (Courant and Hilbert, D.), 228

Minkowski, Hermann

  Born and, 56

  Einstein and, 32–34

  Hilbert, D., and, 18, 19, 24, 25, 38, 65, 272

Minkowski, Hermann *(Continued)*
  invariant theory and, 272
  Lorentz transformation and, 88
  Noether, E., and, 26
  space-time and, 32–34, 65
  special theory of relativity and,
    32–33, 275–276
Möbius strip, 14
Modern Algebra (Waerden), 186–187, 187
    (photo)
modular addition, 288
momentum, 71
  conservation of, 118–119, 126–127,
    138, 140
  energy-momentum four vector,
    121–122
  in Noether's theorem, 289
    moon, crater named after
      Noether on, 299
    Munroe, Randall, 299
    Myspace, 299

naive set theory, 59
nationalism, 79, 93–94
Navier-Stokes equations, 306
Nazis, 151, 154
  antisemitism of, 164–174
  Courant and, 228
  Noether, E., and, 4–5, 261, 297
network science, 266
Neumann, John von, 160, 161–162, 205
neutrons, 133, 291
neutron stars, 124–125
Newton, Isaac, 32
  differential equations of, 71, 117

energy conservation and, 110
gravity and, 44–46, 48, 65, 88, 251
instantaneous action at a distance
  and, 48, 251
laws of motion of, 44, 71, 112, 113, 117,
  118–119, 236
Noether, E. compared to, 247
second law of motion of, 45–46,
  70–71, 113, 117
special theory of relativity and, 123
universal law of gravitation of,
  44–46
Nobel Prize
  of Einstein, 28, 237, 246
  of Laue, 37
  of Lederman, 3
  of Lenard, 165
  Noether, E., and, 246–247
  of Stark, 165
  of Wigner, 224, 226–227
  of Wilczek, 3, 100
Nobel Prize Women in Science
    (McGrayne), 41, 244, 246
Noether, Emmy, 1–2, 10
  abstract algebra and, 63, 131, 189,
    244–245, 285
  antisemitism and, 163–174, 182
  ashes of, 211–214, 214 (photo)
  auditing mathematics classes,
    21–22, 26, 42
  at Bryn Mawr, 178–202, 285
  death of, 197–215
  doctoral thesis of, 27
  Einstein and, 11, 66–67, 74–76, 81,
    150–152, 206–208, 247

encouragement to colleagues and students by, 5–6

energy conservation and, 115

general theory of relativity and, 54–55, 75–76, 81–82, 84–86, 116–117, 131–135, 231, 242

German Mathematicians' Relief Fund and, 195–196

Gordan and, 26–27, 34–36, 42, 243

at Göttingen University, 21–22, 25–26, 42–43, 53–97, 145–174, 203–204, 259–260

graduation of, 26

gravity and, 53–97

habilitation of, 145–163

Hilbert, D., and, 11, 35, 42, 58–59, 75–76, 84, 122, 132, 143, 147–149, 157, 193, 194, 242, 243

hometown of, 16

at Institute for Advanced Study, 189–192, 196–197, 202

invariant theory and, 26, 29, 116, 143, 275

Klein and, 84–86, 115, 122, 143, 147–148, 157, 224–225

preparing to become a language teacher, 20–21

legacy of, 241–263, 296–299

as mathematics teacher, 34–36

memorial service for, 199–200, 202–204, 203 (photo)

memorial stone of, 214, 214 (photo)

methods of working of, 243

Nazis and, 4–5, 261, 297

Newton, compared to, 247

Nobel Prize and, 246–247

obscurity of, 5

peculiarities of, 43, 157–158

recommendation policy of, 193–194

school of mathematics of, 158–159

theory of differential invariants and, 42

at University of Erlangen, 26, 34–36, 241, 275

to US, 4, 175–215

Weyl and, 143, 171, 173, 203–204, 205

youth of, 12–14

Noether, Fritz (brother), 196, 200–202, 205

Noether, Gottfried (nephew), 182, 201, 229

Noether, Hermann (nephew), 201

Noether, Max (father), 26, 34, 35, 42, 164

Bryn Mawr and, 179

invariant theory and, 275

Klein and, 86

Noether Lecture, 297

Noether program, 223

Noether's theorem, 99–143

classical mechanics and, 215, 223, 224, 225, 229, 249

conservation laws in, 100–109, 120, 129–130, 232, 257, 290

control theory and, 307–308

converses and, 108–109, 246

cosmos in, 127, 129–130, 248

econometrics and, 257–258

Eddington and, 236–239

Einstein and, 140–144, 242

Noether's theorem *(Continued)*
elementary particles and, 2
energy conservation in, 117–118, 122, 127–131, 140
energy in, 109–131
extensions of, 296
gauge theories in, 222
general theory of relativity and, 100, 140–144, 217–235, 242, 289
gravity and, 140, 230, 233, 238–239, 250
habilitation and, 150–151
Heisenberg and, 221, 233–234, 235
hidden symmetries and, 289, 291
high-energy physics and, 127, 249
Hilbert, D., and, 231, 242
importance of, 299
Klein and, 242
Lagrangian and, 222–223, 296, 298
legacy of, 241–263
Methods of Mathematical Physics in, 228
momentum conservation in, 140
particle physics and, 224, 225, 231, 249
proof for, 283–284
quantum computing and, 255–256, 300–301
quantum field theory and, 225, 298
quantum mechanics and, 220–221, 225, 233–234, 285, 293, 298
reawakening of, 217–239
representation theory and, 294–296
robotics and, 307–308
scale symmetry and, 253–254
simulations and, 303–306, 304 (fig.), 306 (fig.)
space in, 140
statistical mechanics and, 249
symmetry in, 129–131, 135–140, 226, 235–236, 248–249, 263, 285, 293, 298–299
time in, 140
unified theory and, 2–9, 230
variational calculus and, 233
Veblen and, 221
Weyl and, 220–221
noncommutative property, 295
non-Euclidean geometry, 70, 220, 280
novel, with allusions to Noether, 299

Open University of Israel, 56

Pais, Abraham, 69
parallel postulate, 62, 277–278, 278 (fig.), 279
Park, Marion, 180–181, 182, 199, 208–211
particle physics
grand unified theory and, 250–251
group theory and, 189
Heisenberg and, 120, 223
Noether's theorem and, 224, 225, 231, 249
standard model and, 221–222, 224, 250–251
symmetry in, 223
A Passion for Discovery (Freund), 41
Pauli, Wolfgang (Pauli exclusion principle), 85, 92, 120, 223
Penzias, Arno, 211

perihelion, of Mercury, 95–96, 288, 293–294

Perimeter Institute, 297–298

photons
energy of, 128
in quantum theory, 292

Physics from Symmetry (Schwichtenberg), 233

Planck, Max, 73, 96
Klein and, 225
"Manifesto to the Civilized World" and, 40

plane geometry, 280

Playfair, John, 277

"Plenary Report on Progress in Gauge Theories" (Iliopoulos), 221

Poincaré, Henri, 65, 86–87, 262

point, definition of, 60–61

Pollaczek-Geiringer, Hilda, 154

positrons, 103, 121

postulates (axioms), 61–62, 63

potential energy, 112–113

precession, of Mercury, 44–45, 95–96, 207, 251

prime numbers, Euclid on, 273

Princeton University, 190

principle of equivalence, of Einstein, 46–48

probability, wavefunction and, 292–293

Proceedings of the Royal Göttingen Academy of Science, 143

protons, 89, 103, 136, 292

pulsars, 124–125

Pythagorean theorem, 3, 8, 255

Euclid and, 63

invariant theory and, 270–271

quantum computing, 255–256, 300–301

quantum field theory, 250, 252
Lagrangian and, 222
Noether's theorem and, 225, 298
representation theory and, 295

Quantum Field Theory in a Nutshell (Zee), 298

quantum gravity, 249

quantum mechanics, 48, 221, 252
Born and, 56
Einstein and, 237
on expanding universe, 127–128
group theory and, 189, 292–293
Hermann and, 160, 161–162
Hilbert, D., and, 284–285
invariant theory in, 226
Neumann on, 161
Noether's theorem and, 220–221, 233–234, 293
Paul exclusion principle in, 85
photons in, 292
probabilities in, 160–162
standard model and, 220–221
symmetry in, 122
uncertainty principle in, 221
uniformity of substance in, 139–140
wave function in, 292

qubit, 301

Quest (Infeld), 165

redshift, 128

reference frames, 29–30, 31, 43

Reid, Constance, 9

Relativity, Groups and Topology (summer school), 227

representation theory, 294–296

Reviews of Modern Physics, 225–226, 229

Ricci calculus (tensor calculus), 71–73

Ricci-Curbastro, Gregorio, 71–72

Riemann, Bernhard, 70, 71–72, 279

robotics, 307–308

Rockefeller Foundation, 175, 182–183

Roosevelt, Theodore, 210

Roquette, Peter, 245

Rössler, Otto E., 298–299

rotational symmetry (spatial isotropy), 119, 290

Rowe, David E., 70, 80, 132

Royal Göttingen Academy of Science, Proceedings of the, 143

Runge, Carl, 115, 290, 294

Runge-Kutta method, 115

Runge-Lenz vector, 290, 293–294

Russell, Bertrand, 59

Rust, Berhard, 173–174

SA. *See* Sturmabteilung

Sabin, Florence, 209, 210, 211

Sanayei, Ali, 298–299

scale symmetry, 253–254

Schoenberg, Arnold, 41

Scholar Rescue Fund, of Institute of International Education, 297

Schrödinger, Erwin, 1, 292–293

cat of, 301

quantum mechanics and, 284–285

Schwarzschild, Karl, 26, 96, 207–208

Schwichtenberg, Jakob, 233

Schwinger, Julian, 223

Scott, Charlotte Angas, 179

Shakespeare, William, 113, 167

simulations, in Conway's Game of Life, 303–306, 304 (fig.), 306 (fig.)

Sinclair, Upton, 40

Sitter, Willem de, 92

Social Democratic Party, 172

Sommerfeld, Arnold, 64, 78, 80, 96

Sonnet 154 (Shakespeare), 113

Soviet Union, 196–197, 201–202

space. *See also* curved space

energy-momentum four vector, 121–122

in field theory, 88

in general theory of relativity, 89

Hilbert, 55, 134

Lorentz transformation and, 65, 89

for Newton, 32

in Noether's theorem, 140

in special theory of relativity, 31–32, 120–121

variation of laws of physics in, 139

space-time. *See also* curved space-time

in classical mechanics, 226

energy conservation and, 124

gravitational waves and, 124–125

gravity and, 232, 280–281

Grossmann and, 278–279

Minkowski and, 32–34, 65

in special theory of relativity, 121, 276

symmetry in, 129–130, 221, 226

Space-Time Matter (Weyl), 221, 227
spatial isotropy (rotational symmetry), 119, 290
spatial translation symmetry, 138–139, 289
special theory of relativity, 120–121, 275–276
  of Einstein, 28–33, 43, 69
  Lorentz transformation and, 65
  Newton and, 123
  space-time and, 65
spectrum of hydrogen, 292
speed of light, 30–31
  in $E = mc^2$, 28, 67, 106, 121, 237
  in general theory of relativity, 89
Stalin, Joseph, 197
standard model, 217–235
  conservation laws in, 232
  Courant and, 228–229
  elementary particles in, 236
  gauge theories and, 222
  general acceptance of, 246
  general theory of relativity and, 232
  Glashow and, 222–223
  gravity and, 218, 220, 233
  high-energy physics and, 222, 233
  Hill, E., and, 229–230
  Lagrangian and, 222–223
  magnetic moment of electron in, 219
  particle physics and, 221–222, 224, 250–251
  quantum mechanics and, 220–221
  symmetry in, 232, 250–251

unified theory and, 232
Weyl and, 220–221, 227
Wigner and, 224–227
Stark, Johannes, 165
statistical mechanics
  Einstein and, 249
  Hilbert, D., and, 65
  Noether's theorem and, 249
Stauffer, Ruth, 184
Stein, Edith, 151
Stephens, Ransom, 299
Sternin, Otto, 96
Stewart, Ian, 242
string theory
  gauge theories and, 222
  gravity and, 222
  quantum gravity in, 249
strong force, 133, 232
Sturmabteilung (SA), 169, 173
symmetry
  in abstract algebra, 285–286
  of antimatter twins, 102–103
  in biology, 101–102
  chirality and, 192
  in classical mechanics, 117–118, 226, 229
  and conservation laws, 6–7, 100–109, 120, 121, 127, 129–131, 250–251, 283–284
  defined, 101
  and elementary particles, 226, 236
  and energy conservation, 112, 117–118
  in gauge theories, 221, 222
  in general theory of relativity, 108

symmetry *(Continued)*
  gravity and, 233
  in grids, 303
  group theory and, 188–189
  in Noether's theorem, 100–109,
      117–121, 127, 129–131, 135–140,
      226, 235–236, 248–249, 263,
      285, 293, 298–299
  in particle physics, 223
  in principle of equivalence, 47
  in quantum theory, 122
  rotational, 119, 290
  scale, 253–254
  in space-time, 221, 226
  spatial translation, 138–139, 289
  in special theory of relativity, 276
  in standard model, 232, 250–251
  time translation, 109, 117–118, 122,
      289
  transformations and, 104–105, 117,
      222
  in unified theory, 2, 230

Taussky, Olga, 192–193, 198
Taylor, Joseph, 124–125
tensor calculus (Ricci calculus), 71–73
tensors, 121
theorems, 63. *See also* Noether's
      theorem; Pythagorean
      theorem
  Fermat's last, 68–69
theory of differential invariants, 42
time. *See also* space-time
  Lorentz transformation and, 65, 89
  for Newton, 32, 117

  in Noether's theorem, 140
  in special theory of relativity, 31–32,
      120–121
time translation symmetry, 109,
      117–118, 122, 289
time zero, 129
Topologie (Alexandrov and Hopf), 245
Tower of Pisa experiment, of Galileo,
      46
Transactions, of American
      Mathematical Society, 179
transformations
  gauge, 222
  in symmetry, 104–105, 117, 222
Twain, Mark, 9

uncertainty principle, 221
Underwood, Thomas, 185
unification, Newton's law of gravity
      as 44
unified theory, 250. *See also* grand
      unified theory
  Hilbert, D., and, 133–134, 231
  Maxell's electromagnetic theory
      and, 31, 88–90
  Noether's theorem and, 2–9, 230
  standard model and, 232
uniformity of substance, 139–140
universal theory of gravitation, of
      Newton, 44–46
University of Erlangen, 25
  fiftieth anniversary
      commemoration at, 241
  Klein at, 86, 164
  Noether, E., at, 26, 34–36, 241, 275

"The Unreasonable Effectiveness of Mathematics in the Natural Sciences" (Wigner), 226–227

variational calculus, 131, 228, 233
variational principle and general relativity, 92–93
Veblen, Oswald, 178–179, 190, 191, 221
vectors, 72–73
  angular momentum and, 290
  energy-momentum four vector, 121–122
  invariance of, 76
  Runge-Lenz, 290, 293–294
Voigt, Woldemar, 65

Waerden, B. L. van der, 159, 186–188, 187 (photo), 189
wave function in quantum mechanics, 292
wave mechanics, 284
weak force, 232
Weil, André, 205
Weinberg, Steven, 298
Weyl, Hermann, 8, 70
  antisemitism and, 171–172
  Einstein and, 91–92, 133, 220
  German Mathematicians' Relief Fund and, 195–196
  at Göttingen University, 23–24, 203–204
  Hilbert, D., and, 23–24, 220, 274

  at Institute for Advanced Study, 171–172, 194
  Noether, E., and, 22, 27, 143, 171, 173, 203–204, 205
  Noether, F., and, 200, 201
  Noether's theorem and, 220–221
  representation theory and, 295
  standard model and, 220–221, 227
Wheeler, Anna Pell, 185, 186, 190, 195, 198
Wheeler, John, 122, 179
Whetstone, Ethel, 213–214
Whores, Guild of, 48
Wiener, Norbert, 176
Wigner, Eugene, 224–227
Wilczek, Frank, 3, 100, 139–140
Wilde, Oscar, 259
Wilson, Robert, 211
Wolfskehl, Paul (Wolfskehl Prize), 68–69
World War I
  antisemitism after, 165
  "Manifesto to the Civilized World" of, 40, 79
  Schwartzschild in, 207–208
wormholes, 96

XKCD, 299

Young, Grace Chisholm, 151, 262
Young, Laurence, 262

Zangger, Heinrich, 77–79, 81, 82
Zee, Anthony, 298

Credit: Cinthya Lara

LEE PHILLIPS grew up on the Lower East Side of Manhattan. He attended Stuyvesant High School, Hampshire College, and Dartmouth College, where he received a PhD in theoretical physics. He was a Research Physicist at the US Naval Research Laboratory for twenty-two years, working on projects for the Navy, the National Aeronautics and Space Administration, the Department of Energy, and other government agencies. He has written many popular articles on science, technology, and their history and is involved with science education and outreach, including serving on the Board of Directors of the Friends of Arlington's Planetarium. Phillips lives in Honduras, where he writes and conducts research while studying Spanish.